PHYSICS RESEARCH AND TECHNOLOGY

AN ESSENTIAL GUIDE TO ELECTRICAL CONDUCTIVITY AND RESISTIVITY

PHYSICS RESEARCH AND TECHNOLOGY

Additional books and e-books in this series can be found
on Nova's website under the Series tab.

PHYSICS RESEARCH AND TECHNOLOGY

AN ESSENTIAL GUIDE TO ELECTRICAL CONDUCTIVITY AND RESISTIVITY

LUKE LEWIN
EDITOR

nova
science publishers
New York

NOTICE TO THE READER

Library of Congress Cataloging-in-Publication Data

ISBN: 978-1-53615-047-6

Published by Nova Science Publishers, Inc. † New York

CONTENTS

PREFACE

An Essential Guide to Electrical Conductivity and Resistivity opens with experimental and theoretical data on the important structurally sensitive property of the molten oxide-chloride systems KCl (50 mol. %)–PbCl2 (50 mol. %), CsCl (18.3 mol. %)–PbCl2 (81.7 mol. %) and CsCl (71.3 mol. %)–PbCl2 (28.7 mol. %) with PbO concentration reaching 20 mol.% in the temperature range of 764 – 917 K.

This book also reports on recent developments in the stabilization, carbonization, and activation of naturally grown biomass, their physical and chemical properties, and major applications in supercapacitors.

A brief overview of the application areas of the electrical resistivity method, a non-destructive and inexpensive geophysical technique, as well as comprehensive and practical overview of its results in environmental and geotechnical projects, archeology and stone cultural heritage is presented.

Chapter 1 - Experimental and theoretical data on the important structurally sensitive property of the molten oxide-chloride systems KCl (50 mol.%)–$PbCl_2$ (50 mol.%), CsCl (18.3 mol.%)–$PbCl_2$ (81.7 mol.%) and CsCl (71.3 mol.%)–$PbCl_2$ (28.7 mol.%) with PbO concentration reaching 20 mol.% in the temperature range of 764 – 917 K are analyzed and generalized. The measuring technique for the oxy-chloride melts electrical conductivity is developed. It implies continuous measuring of the impedance of the cell with parallel lead electrodes after the lead oxide addition to the melt and considers the cell constant temperature dependence, which allows performing more accurate measurements of electric conductivity in the wide temperature range. The influence of cations on the oxy-chloride melt electrical conductivity is analyzed. The values of the electrical conductivity activation energy (E_κ) are calculated and the interaction mechanism between lead oxide and chlorides in the systems under study is suggested.

Chapter 2 - Activated carbon (AC) is carbon that is processed at elevated temperatures with activating agents. It has relatively low-volume pores, meaning increased surface area, in order to store a considerable amount of energy. In addition to coal and coke, AC can also be produced from natural materials, which can be inexpensive and sustainable natural sources for utilization in energy storage devices, such as supercapacitors and battery anodes. Many nutshells, coconut shell, bamboo, sugarcane bagasse, rice bran, corn cob, and potato wastes can be good sources of AC for different applications. These natural materials are used as a precursor, carbonized at high temperature in an oxygen free environment, and activated by physical or chemical activation processes to produce AC with a very high specific surface area ($\sim$1500 m^2/g). The activated carbon is used to fabricate electrodes of supercapacitors and is characterized using different techniques prior to the application. These supercapacitors have a high specific capacitance, high energy density, and high power density compared to many other storage devices. This book chapter reports on recent developments in stabilization, carbonization, and activation of naturally grown biomass; their physical and chemical properties; and major applications in supercapacitors. This study shows that natural materials are promising options for producing ACs for different supercapacitor applications at a low cost. Many readers, such as students, engineers, scientists, and other participants in supercapacitor technologies, will greatly benefit from this work.

Chapter 3 - The electrical resistivity method is a non-destructive and inexpensive geophysical technique that can be applied to a variety of geological targets at small and medium depths. The method allows a mapping of the internal structure of investigated medium. Initially applied to mining and groundwater exploration, new applications related to environmental and geotechnical projects, archeology and stone cultural heritage have emerged in the last three decades. The method is also widely used in downhole logging. This chapter presents a brief overview of application areas of this geophysical method and gives a comprehensive and practical overview of the results obtained in each of these areas.

Chapter 1

ELECTRICAL CONDUCTIVITY OF MOLTEN MIXTURES OF LEAD CHLORIDE AND LEAD OXIDE CONTAINING POTASSIUM AND CESIUM CHLORIDES

Pavel A. Arkhipov
Institute of High Temperature Electrochemistry,
Ural Branch of Russian Academy of Sciences,
Ekaterinburg, Russian Federation

ABSTRACT

Experimental and theoretical data on the important structurally sensitive property of the molten oxide-chloride systems KCl (50 mol.%)–$PbCl_2$ (50 mol.%), CsCl (18.3 mol.%)–$PbCl_2$ (81.7 mol.%) and CsCl (71.3 mol.%)–$PbCl_2$ (28.7 mol.%) with PbO concentration reaching 20 mol.% in the temperature range of 764 – 917 K are analyzed and generalized. The measuring technique for the oxy-chloride melts electrical conductivity is developed. It implies continuous measuring of the impedance of the cell with parallel lead electrodes after the lead oxide addition to the melt and considers the cell constant temperature dependence, which allows performing more accurate measurements of electric conductivity in the wide temperature range. The influence of cations on the oxy-chloride melt electrical conductivity is analyzed. The values of the electrical conductivity activation energy (E_κ) are calculated and the interaction mechanism between lead oxide and chlorides in the systems under study is suggested.

INTRODUCTION

Electrical conductivity (κ) of molten systems is important from both theoretical and practical approaches, based on structurally sensitive property that provides information on the nature of current transporting particles and their mobility [1]. Fundamental research on electrical conductivity as well as on other physical chemical properties may significantly elucidate structural peculiarities of molten salts. In addition, electrical conductivity affects production technological parameters, determines the electrolytic cell thermal balance and influences power consumption during electrochemical processes.

MEASURING METHOD OF ELECTRICAL CONDUCTIVITY

It is a complicated experimental task to measure the electrical conductivity of oxy-chloride melts. Apart from production difficulties, which appear at such studies, including the dependence of the electrolyte resistance on the alternating current frequency and etc., high corrosion activity of oxy-chloride melts, lead oxide dissolution dynamics in chloride mixture and molten mixture evaporation should be considered.

CELLS WITH TWO PARALLEL ELECTRODES

To measure electrical conductivity the cells with parallel electrodes, which are more convenient under the present conditions as compared to capillary cells [2, 3], were used. Cells with parallel electrodes allow recording changes in the melt electrical conductivity at a gradual oxide addition in a single experiment; measuring electrolytes electrical conductivity in a wide temperature range, including a two-phase region; operating with molten oxy-chloride salts during a relatively short period of time.

Resistance of molten salts is measured by an alternating current bridge [4] or by impedance meters [5, 6]. Impedance spectroscopy is the most modern and accurate method to measure the melts resistance. However, it requires accurate equivalent schemes for various electrochemical systems.

In papers [7-10] resistance of molten salts was measured by the method of "continuously changing cell constant," which allowed researchers to decrease the data scattering. This method is based on the principle of a continuous change in the cell constant by varying the immersion depth of the platinum electrode in pyrolytic boron nitride tube into the melt. In the present work the electrical conductivity of lead and cesium chlorides based molten mixtures was determined in cells with two liquid metal electrodes [11, 12-15].

Figure 1 represents a scheme of the experimental cell. An alundum crucible (2) with the electrolyte under study (8) was placed into the isothermal area of the shaft-type resistance furnace with nichrome heaters, the temperature inside was controlled by a microprocessor thermoregulator VARTA TP-403. When the electrolyte (8) melted a measuring set up with liquid metal electrodes was immersed into the melt. The set up was made of parallel alundum covers (3) with similar openings for the lead electrodes (1) interconnection with the electrolyte. Molybdenum bars of 1 mm in diameter (5) are protected from the electrolyte contact by alundum tubes (4) and provide lead contact with the measuring set up. The melt temperature was measured by a Pt/Pt-Rh thermocouple (6), which was placed in immediate proximity to the electrodes and was protected from the melt by an alundum cover (7). Lead oxide was added to the melt through a special device.

After each lead oxide addition, the electrolyte resistance was recorded. During the experiment the electrodes location remained unchanged. The

                          Pavel A. Arkhipov

immersion depth of the measuring electrodes was constant during the experiment.

The molten mixture resistance was measured by a galvanostate - potentiostate Avtolab 302N (Holland). The impedance diagrams were recorded within the alternating current frequency interval from 100000 to1000 Hz with the current amplitude of 10 mA.

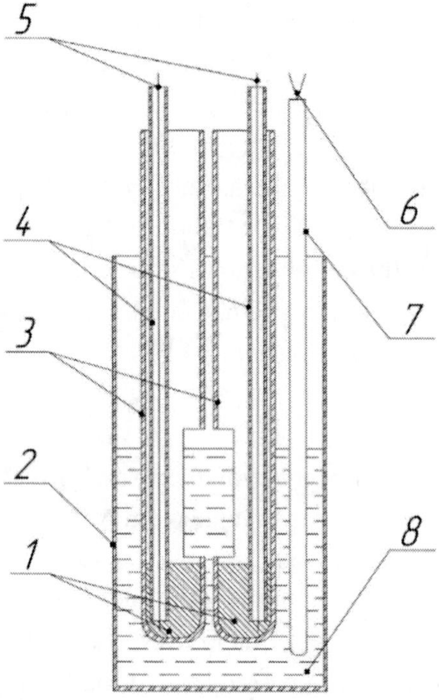

Figure 1. Schematic of the cell for the electrical conductivity measurement. 1 – C1 grade lead; 2 – Al_2O_3 crucible; 3 – Al_2O_3 covers with openings; 4 – Al_2O_3 tubes; 5 – current leads; 6 – Pt/Pt-Rh thermocouple; 7 – Al_2O_3 cover for the thermocouple; 8 – melt.

The electrochemical cell impedance (Z) includes the electrodes impedance, which is determined by the reactive compound $Z''(\Omega)$, and the electrolyte resistance, which is determined by the active compound $Z'(\Omega)$. A computer-aided system of impedance measurements Avtolab 302N records both active and reactive compounds. The electrolyte resistance was detected via the impedance diagrams, i.e., according to the value of the real impedance region at the intersection of the curve with the X-axis $Z''(\Omega) = 0$. Figure 2 presents the hodograph diagrams for the CsCl (73.6 mol.%) - PbCl$_2$ (26.4 mol.%) melt at the moment of the PbO (15 mol.%) addition at the temperature of 765 K.

Diagrams illustrate changes in the cell impedance after the PbO addition. The impedance value was assumed to be the Y-axis at zero value of the X-axis, when the resistance was constant.

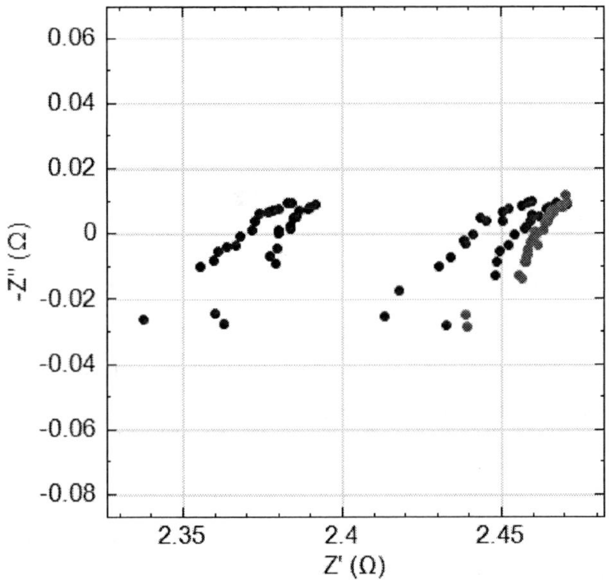

Figure 2. Impedance diagram of the CsCl(73.6 mol.%)-PbCl$_2$ (26.4 mol.%) + PbO (15 mol.%) melt.

The melt resistance, which was determined via impedance diagrams, was used to calculate the value of electrical conductivity. The measuring set-up was calibrated in the CsCl (73.6 mol.%)–PbCl$_2$ (26.4 mol.%) melt. The measurements were performed in the temperature interval of 764 -917 K. The temperature was recorded by a digital multimeter APPA 109N with the frequency of 1 measurement per second. We used explicit data on the electrical conductivity of this melt reported by Janz [16]. The specific electrical conductivity was calculated using the formula $\kappa = K/R$, where K is the cell constant, R is the electrolyte resistance.

The electrical conductivity of each melt was measured at the temperatures above the liquidus. The electrical conductivity of the electrolytes under study in the temperature interval of 764 – 917 K was calculated considering the temperature dependence of the cell constant, which value is $1.4 - 1.6 \ m^{-1} \cdot 10^{-2}$.

ELECTRICAL CONDUCTIVITY
OF KCL-PBCL$_2$-PBO MELTS

Studies on the electrical conductivity of systems, which are of interest for obtaining and refining heavy low-melting metals, are reported in papers [17-22]. Tarasova [23] and Bloom and Heymann [24] present interesting data on the electrical conductivity of the KCl-PbCl$_2$ system in the concentration interval of 0-35 mol.% and the temperature interval of 460 - 720 °C. The most definitive values of the electrical conductivity were provided by Lantratov et al. [25]. The authors studied the electrical conductivity of the PbCl$_2$-KCl systems in the KCl concentration interval of 0 -100% in the temperature interval of 425- 800°C. The values of electrical conductivity for the PbCl$_2$-KCl eutectic melt are 0.937 S/(m$\cdot$10^{-2}) at 450°C and 1.716 S/(m$\cdot$10^{-2}) at 700°C. There is few published data on the study of the oxide additions influence on the electrical conductivity of halide melts. In the KF–KCl–K$_2$SiF$_6$–SiO$_2$ system [26] the 4 mol.% addition of SiO$_2$ at the temperature of 738 °C decreased the specific electrical conductivity by 7% as compared to that of the KF–KCl–K$_2$SiF$_6$ melt. The addition of 40 mol.% of SiO$_2$ to the Na$_3$AlF$_6$-AlF$_3$ melt decreased the electrical conductivity by more than 30% [27]. The addition of tungsten oxide decreased the NaF-NaCl melt electrical conductivity [28]. For instance, at the addition of 25 mol.% of WO$_3$ at 950 °C the electrical conductivity of the fluoride-chloride melt changed from 2.15 to 1.94 S/(m$\cdot$10^{-2}). The author explained that this process is caused by the formation of complex ions. The increase in the WO$_3$ concentration results in the decrease in the NaF concentration because of the formation of additional anionic complex groupings $WO_3F_3^{3-}$, which leads to the decrease in the total number of ions, which take part in the charge transport [28]. The influence of aluminium oxide on the electrical conductivity of molten cryolites was analyzed in papers [29-33]. Generally, the increase in the Al_2O_3 concentration decreased the cryolites electrical conductivity. For example, the increase in the aluminium oxide concentration by 1 mol.% decreased the specific electrical conductivity of the $(Na_3AlF_6$-40%$K_3AlF_6)$-AlF_3-Al_2O_3 melt by 0.07 $S/(m \cdot 10^{-2})$ [33]. There is a lack of data on the influence of lead oxide on the chloride melts electrical conductivity.

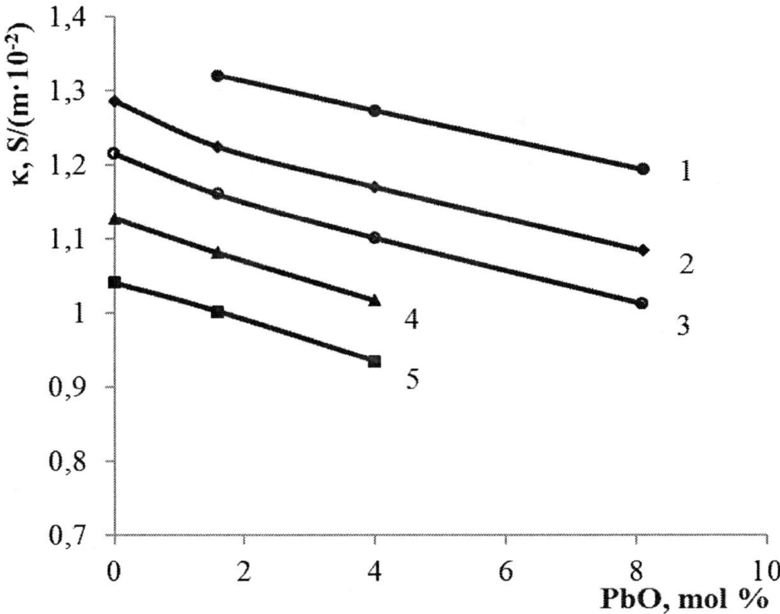

Figure 3. Electrical conductivity of the $PbCl_2$ – KCl – PbO melt at the temperature K: 1 – 853; 2 – 823; 3 – 803; 4 – 778; 5 – 753.

The electrical conductivity of the KCl-$PbCl_2$-PbO system with additions of 1.59, 3.99 and 8.07 mol.% of lead oxide was studied in the cell with parallel electrodes in the temperature interval of 700 – 850 K [34-38]. The results of the electrical conductivity measurements are presented in the form of concentration dependencies (see Figure 3).

The value of electrical conductivity for the $PbCl_2$-KCl eutectic melt is 1.041 S/(m·10^{-2}) at 753 K and 1.36 S/(m·10^{-2}) at 853 K. At the addition of 1.59 mol.% of lead oxide to the $PbCl_2$ – KCl chloride system the melt electrical conductivity decreases to the values of 1.002 S/(m·10^{-2}) at 723 K and 1.32 S/(m·10^{-2}) at 853 K. A further increase in the PbO concentration to 3.99 mol.% in the molten mixture decreases the value of electrical conductivity to 1.273 S/(m·10^{-2}) at the temperature of 853 K. The addition of 8.07 mol.% of PbO at the temperature of 853 K decreases the electrical conductivity by 15% as compared to that of the $PbCl_2$-KCl (50:50 mol.%) melt.

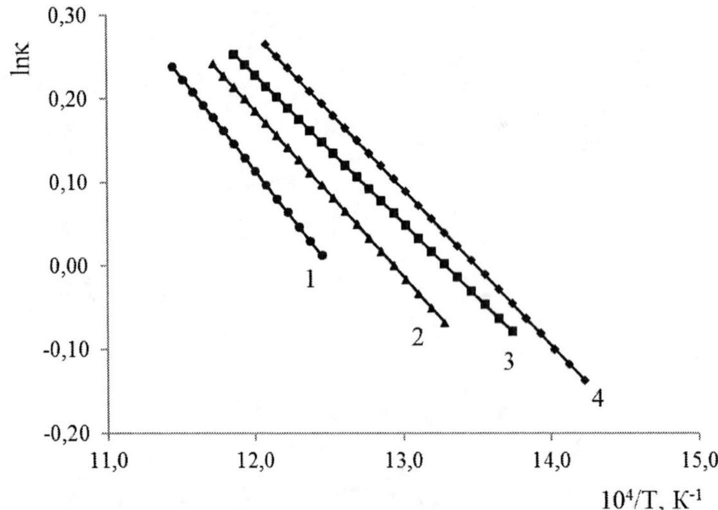

Figure 4. Electrical conductivity of the $PbCl_2$ – KCl – PbO melt at the PbO concentration, mol.%: 1 – 8.07; 2 – 3.99; 3 – 1.59; 4 – 0.

On the κ temperature experimental curves in the «ln κ on 1/T» coordinates (Figure 4) the κ ln functions decrease linearly as the 1/T argument increases. The inclination of conductivity polytherms in these coordinates for all studied compositions does not depend on temperature.

Therefore, the electrical conductivity behavior in each melt is characterized by a single value of the electrical conductivity activation energy in the whole temperature interval. The values of the electrical conductivity activation energy, which were found according to the inclination angle tangent $E_\kappa = R \cdot tg\alpha$, are presented in Table 1.

Table 1. Electrical conductivity and electrical conductivity activation energy of PbCl$_2$-KCl (1:1) with PbO addition

PbO concentration, mol.%	κ, S/(m·10^{-2})			E_κ, kJ/mol.	Temperature interval, K
	T = 753 K	T = 803 K	T = 853 K		
1.59	1.002	1.160	1.321	14.1	728-853
3.99	0.934	1.101	1.273	16.5	748-853
8.07	-	1.013	1.194	18.8	803-873

The values of electrical conductivity activation energy (E_κ) of the (PbCl$_2$ – KCl) – PbO molten mixtures increase as the lead oxide concentration increases. This peculiarity elucidates structural changes in the oxy-chloride system towards the formation of more complex ionic groupings, such as $[Cl_2Pb\text{-}O\text{-}PbCl_2]^{2-}$.

ELECTRICAL CONDUCTIVITY OF CsCl-PbCl$_2$-PbO MELTS

To perform a more detailed study of the transport mechanism of lead oxide and chloride containing oxy-chloride melts ionic conductivity the CsCl-PbCl$_2$ chloride melt was chosen as a model system. The electrical conductivity was measured in the compositions of cesium and lead chlorides, which correspond to the following eutectics: CsCl (18.3 mol.%) – PbCl$_2$ (81.7 mol.%) and CsCl (71.3 mol.%) – PbCl$_2$ (28.7 mol.%) with PbO additions [9-12]. Insignificant dissolution of lead oxide was observed in these systems [13]. Another peculiarity is that the Cs$^+$ cation, which is characterized by small ionic potential and weak resistance to the Pb^{2+} cation polarization, may form stable complex ionic lead-containing groupings.

Pavel A. Arkhipov

ELECTRICAL CONDUCTIVITY OF THE CsCl (18.3 MOL.%)–PbCl₂ (81.7 MOL.%) MELT

The electrical conductivity of the CsCl (18.3 mol.%)–$PbCl_2$ (81.7 mol.%) system with lead oxide additions ranging from 0 to 18 mol.% was studied in the cell with parallel electrodes in the temperature interval of 723 – 863 K [9-12]. Figure 5 illustrates the concentration dependencies of the electrical conductivity.

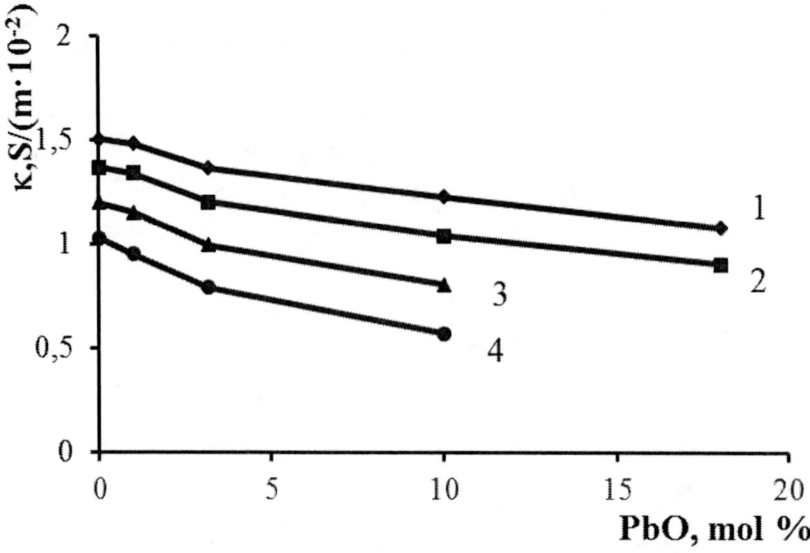

Figure 5. Electrical conductivity of the CsCl (18.3 mol.%)–$PbCl_2$ (81.7 mol.%) – PbO melt at the temperature, K: 1 – 863; 2 – 823; 3 – 773; 4 – 723.

The values of electrical conductivity measured for the eutectic CsCl (18.3 mol.%)–$PbCl_2$ (81.7 mol.%) melt are 1.027 S/(m·10⁻²) at 723 K and 1.503 S/(m·10⁻²) at 863 K, which is in a good agreement with literature data [16]. The addition of 3.2 mol.% of lead oxide to the chloride mixture decreases the electrical conductivity to the values of 0.789 S/(m·10⁻²) at 723 K and 1.363 S/(m·10⁻²) at 863K. The further increase in the PbO concentration to 10 mol.% in the molten mixture decreases the electrical conductivity value to 1.229 S/(m·10⁻²) at the temperature of 863 K. The addition of 18 mol.% of PbO at the temperature of 863 K decreases the electrical conductivity by 28% as compared to that of the CsCl (18.3 mol.%)–PbCl₂ (81.7 mol.%) melt.

Figure 6 demonstrates experimental temperature dependencies of κ in the «ln κ on 1/T» coordinates. The κ ln functions decrease as the 1/T argument grows. The polythermal electrical conductivity inclination in these coordinates, for all studied compositions, does not depend on temperature.

Table 2 provides the values of specific electrical conductivity activation energy, which were calculated according to the inclination angle tangent.

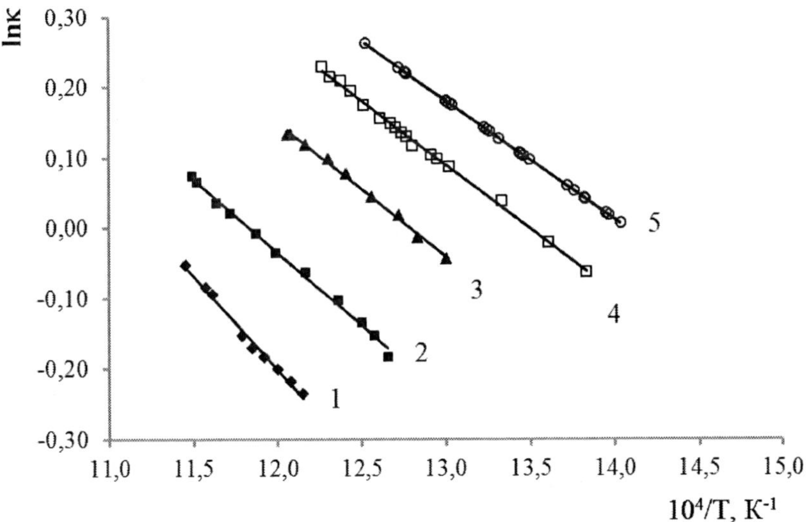

Figure 6. Electrical conductivity of the CsCl (18.3 mol.%)–PbCl₂ (81.7 mol.%) – PbO melt at the PbO concentration, mol.%:1 – 18,0; 2 – 10,0; 3 – 3,2; 4 – 1,0; 5 – 0.

The values of the electrical conductivity activation energy (E_κ) in the CsCl (18.3 mol.%)–PbCl₂ (81.7 mol.%) – PbO molten system increase as the lead oxide concentration increases. The same tendency was observed in the (PbCl₂ – KCl) – PbO system, which testifies the identical structural changes in ionic melts.

Pavel A. Arkhipov

Table 2. Values of electrical conductivity and electrical conductivity activation energy of the CsCl (18.3 mol.%)–PbCl$_2$ (81.7 mol.%) melt with PbO additions

PbO concentration, mol.%	κ, S/(m·10-2)			Eκ, kJ/mol.	Temperature interval, К
	T = 773 K	T = 823 K	T = 873 K		
1.0	1.104	1.285	-	15.2	723-838
3.2	0.977	1.127	-	15.7	769-829
10.0	0.769	0.940	1.180	17.3	790-873
18.0	-	0.791	0.950	22.2	813-873

ELECTRICAL CONDUCTIVITY OF THE CSCL (71.3 MOL.%)–PBCL$_2$ (28.7 MOL.%) MELT

Electrical conductivity of the CsCl (71.3 mol.%)–PbCl$_2$ (28.7 mol.%) melt with lead oxide additions ranging from 0 to 15 mol.% was studied in the cell with parallel electrodes in the temperature range of 757 – 917 K [9-12]. Figure 7 illustrates the concentration dependencies of electrical conductivity.

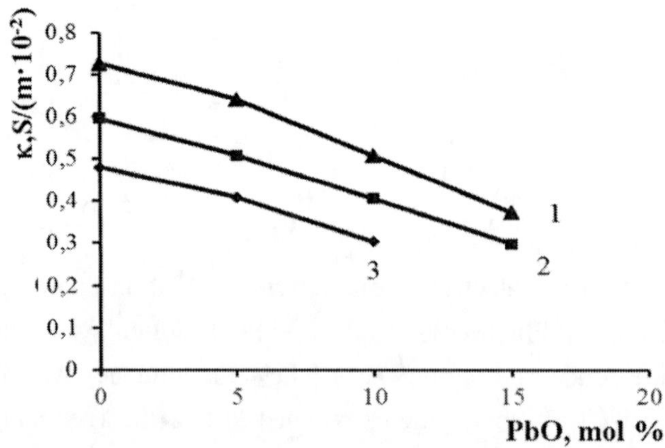

Figure 7. Electrical conductivity of the CsCl (71.3 mol.%)–PbCl$_2$ (28.7 mol.%) – PbO melt at the temperature, K: 1 – 873; 2 – 823; 3 – 773.

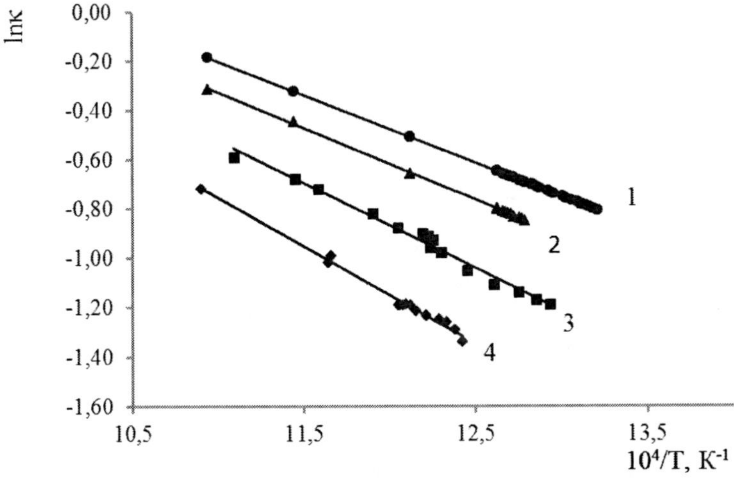

Figure 8. Electrical conductivity of the CsCl (71.3 mol.%)–PbCl$_2$ (28.7 mol.%) – PbO melt at the PbO concentration, mol.%:1 – 0.0; 2 – 5.0; 3 – 10.0; 4 – 15.0.

Table 3. Values of electrical conductivity and electrical conductivity activation energy of the CsCl (71.3 mol.%)–PbCl$_2$ (28.7 mol.%) – PbO with PbO additons

PbO concentration, mol.%	κ, S/(m·10-2)			Eκ, kJ/mol.	Temperature interval, К
	T = 773 K	T = 823 K	T = 873 K		
0.0	0.48	0.596	0.725	22.8	757-914
5.0	0.409	0.508	0.641	24.2	773-914
10.0	0.304	0.405	0.507	28.7	773-901
15.0	-	0.298	0.372	39.7	805-917

The obtained values of electrical conductivity for the eutectic CsCl (71.3 mol.%)–PbCl$_2$ (28.7 mol.%) melt are equal to 0.48 S/(m·10^{-2}) at 773 K and 0.725 S/(m·10^{-2}) at 873 K, which is well within one confidence interval with literature data [16]. The addition of 5.0 mol.% of lead oxide to this chloride mixture decreases the electrical conductivity of the melt to the values of 0.409 S/(m·10^{-2}) at 773 K and 0.641 S/(m·10^{-2}) at 873 K. The further increase in the PbO concertation to 10 mol.% in the molten mixture

                    Pavel A. Arkhipov

decreases the value of electrical conductivity to 0.507 S/(m·10^{-2}) at the temperature of 873 K. The addition of 15 mol.% of PbO at the temperature of 873 K decreases the electrical conductivity by 48% as compared to that of the CsCl (71.3 mol.%)–PbCl$_2$ (28.7 mol.%) melt.

Figure 8 presents the experimental temperature dependencies of κ in the ln κ on $1/T$ coordinates. The κ ln functions decrease as the 1/T argument increases.

The inclination of the polythermal electrical conductivity in these coordinates, for all compositions under study, does not depend on temperature. Therefore, the electrical conductivity behavior of each melt is characterized by a single in the whole temperature range value of the electrical conductivity activation energy. Table 3 demonstrates the values of the specific conductivity activation energy, which were calculated according to the inclination angle tangent $E_κ = R \cdot tgα$.

The calculated values of electrical conductivity activation energy ($E_κ$) of the CsCl (18.3 mol.%) – PbCl$_2$ (81.7 mol.%) – PbO molten system are nearly twice greater than the $E_κ$ values obtained for the PbCl$_2$ – KCl – PbO melt in the temperature interval under study and similar lead oxide concentration in the melts. Such decrease in the electrical conductivity activation energy, when potassium chloride was substituted by cesium chloride, was observed at the studies of ionic conductivity of binary mixtures with rare-earth metal chlorides [39-44].

INFLUENCE OF CATION COMPOSITION ON THE ELECTRICAL CONDUCTIVITY OF OXY-CHLORIDE MELTS

Pure molten potassium, cesium and lead chlorides are considered to be highly conducting ionic compounds [16, 45]. A decrease in the electrical conductivity, which is accompanied by the cation size increase, testifies that the current transport in the melt is mainly realized by cations. Additive changes in the electrical conductivity of chloride mixtures of different compositions were not observed in practical work. Table 4 presents the generalized data on the electrical conductivity of oxy-chloride systems with different cation compositions.

Table 4. Values of electrical conductivity of the KCl–PbCl$_2$–PbO and CsCl–PbCl$_2$–PbO melts at the temperature of 823 K

N	Melt composition	κ, S/(m·10-2)		
		PbO, mol.%		
		1	5	10
1	KCl (50 mol.%) – PbCl2(50 mol.%)	1.252	1.120	1.071
2	CsCl (18.3 mol.%)–PbCl2 (81.7 mol.%)	1.340	1.152	1.041
3	CsCl (71.3 mol.%)–PbCl2 (28.7 mol.%)	0.578	0.508	0.405

Experimental data demonstrate that at the lead oxide concertation in the melt ranging from 1.0 to 5.0 mol.% the specific electrical conductivity increases as the Pb^{2+} concentration increases and the alkali metal cation concentration in the molten oxy-chloride mixture decreases. In the melts with a large Cs^+ (composition 3) concentration the electrical conductivity value is more than twice smaller than that in the melts with a large Pb^{2+} (composition 2) concentration. The common tendency, that the electrical conductivity of the melt decreases as the lead oxide concentration in the melt grows, is true for all studied compositions. The recorded changes in the electrical conductivity of chloride and oxy-chloride melts are associated with the peculiarities of their structures and may be explained by changes in the electrolyte ionic composition, which were observed at the Raman spectroscopy analysis [14]. Indeed, the destruction of $PbCl_3^-$ complex anions and increase in the Coulomb interaction between particles in the molten chloride of the CsCl (18.3 mol.%) – PbCl$_2$ (81.7 mol.%) composition as compared to that in the melt of the CsCl (71.3 mol.%) – PbCl$_2$ (28.7 mol.%) composition result in the relative increase in the electrical conductivity under isothermal conditions. A significant degradation of ion-conducting properties of the PbO - PbCl$_2$ – CsCl oxy-chloride melts as compared to that of the PbCl$_2$ – CsCl chloride electrolytes is associated with the presence of $Pb_3O_2Cl^+$ groupings of the mixed oxy-chloride composition. The formation of the complex oxy-chloride grouping $Pb_3O_2Cl^+$ at addition of lead oxide

                    Pavel A. Arkhipov

into the $PbCl_2$ and $CsCl$ molten salt mixture causes the following changes: a part of Pb^{2+}cations and Cl^- anions stop participating in the electrotransport process and thus the electrical conductivity of the melt decreases.

REFERENCES

[1] Yesin, O. A. Physical chemistry of pyrometallurgical processes. O. A. Yesin, P. V. Gueld, part 2, M.: *Metallurgiya*, 1966, 520p. (in Russian).

[2] Kryukovsky, V. Electrical conductivity of low melting cryolite melts. V. Kryukovsky, A. Frolov, O. Tkacheva, A. Redkin, Yu. Zaikov, V. Khokhlov, A. Apisarov. *Light Metals*. - 2006. - P. 409.

[3] Apisarov, A. P. Electrical conductivity of cryolite-alumina melts with LiF and KF additions. A. P. Apisarov, O. Yu. Tkacheva, Yu. P. Zaykov, N. G. Molchanova. *The melts*. – 20066. - №4. – p. 45. (in Russian).

[4] Smirnov, M. V. Electrical conductivity of molten fluoride alkali metals. M. V. Smirnov, Yu. A. Shumov, V. A. Khokhlov. *Electrochemistry of molten and solid electrolytes*. Works of the Electrochemistry Institute, Sverdlovsk. - 1972. - Edition. 18. -p. 3. (in Russian).

[5] Hives, J. Electrical conductivity of molten cryolite-based mixtures obtained with a tube-type cell made of pyrolytic boron nitride. J. Hives, J. Thonstad. *Light Metals*. - 1994. P. 187.

[6] Wang, L. The electrical conductivity of cryolite melts containing aluminum carbide. L. Wang, A. T. Tabereaux, N. E. Richards. *Light Metals*. - 1994. P. 177.

[7] Wang, X. Electrical conductivity of cryolite melts / X. Wang, R. D. Peterson, T. Tabereaux. *Light Metals*. -1992. – P. 481.

[8] Huang, Y. Electrical conductivity of $(Na_3AlF_6\text{-}40$ wt.%$K_3AlF_6)\text{-}AlF_3$ melts / Y. Huang, Z. Lai, J. Tian, J. Li, Y. Liu. *Light Metals*. -2008. – P. 519.

[9]   Wang, X. A multiple regression equation for the electrical conductivity of cryolite melts / X. Wang, R. D. Peterson, T. Tabereaux. *Light Metals*. - 1993. – P. 247.

[10] Yang, J. Conductivity of KF-NaF-AlF$_3$ system low-temperature electrolyte / J. Yang, W. Li, H. Yan, D. Liu. *Light Metals*. – 2013. – P. 689.

[11] Zakir'yanova I. D. Raman spectra and conductivity of PbO-PbCl$_2$-CsCl melts / I. D. Zakir'yanova, P. A. Arkhipov – *The melts*. – 2016. – № 5. – pp. 404-412. (in Russian).

[12] Arkhipov, P. A. Electrical Conductivity of the CsCl-PbCl$_2$-PbO System / P. A. Arkhipov, A. S. Kholkina, I. D. Zakiryanova, A. V. Bausheva, A. O. Khudorozhkova. *Journal of The Electrochemical Society.* – 2016. – V. 163. – № 10. – P. H881-H883.

[13] Zakir'yanova I. D. Physical-chemical properties of the PbO-PbCl$_2$-CsCl system / I. D. Zakir'yanova, P. A. Arkhipov, I. V. Korzun, D. O. Zakir'yanov, A. S. Kholkina, A. V. Bausheva, A. O. Khudorozhkova. *XX Mendeleyev conference on general and applied chemistry. Ekaterinburg, 26-30 of September 2016. Proceedings.* – 2016. – V. 1 – P. 190. (in Russian).

[14] Zakir'yanova, I. D. Raman Spectra and Conductivity of PbO–PbCl$_2$–CsCl Melts / I. D. Zakir'yanova, P. A. Arkhipov - *Russian Metallurgy (Metally).* – Vol. 2017. – No. 2. – P. 86–90 (перевод статьи из журнала: Расплавы. 2016. № 5. С. 404-412).

[15] Kholkina A. S. Electrical conductivity of the molten CsCl-PlCl$_2$-PbO system / A. S. Kholkina, P. A. Arkhipov, A. V. Bausheva, A. O. Khudorozhkova. Problems of theoretical and experimental chemistry. *Proceedings of the 26th Russian Youth Scientific Conference.* Ekaterinburg. 27-29 of April 2016. Ekaterinburg: Publishing house Of the Ural University. – 2016. – pp. 374-375. (in Russain).

[16] Janz, J. Thermodynamic and Transport Properties for Molten Salts: Correlation Equations for Critically Evaluated Density, Surface Tension, Electrical Conductance, and Viscosity Data / J. Janz. *J. Phys. and Chem. Ref. Data.* – 1988. – V. 17. – N. 2. – P. 1-309.

[17] Duke, F. R. Transport Numbers and Ionic Mobilities in the System Potassium Chloride-Lead Chloride / F. R. Duke, R. A. Fleming. *J. Electrochem. Soc.* – 1957. – v. 104. – P. 251 – 254.

[18] Zhiyuan, Chen Electrical conductivity of $CaCl_2$–KCl–NaCl system at 1080 K / Chen Zhiyuan, Liu Junhao, Yu Ziyou, Chou Kuo-Chih. *Thermochimica Acta.* –2012. – 543. – P. 107–112.

[19] Rafalskiy, V. V. Physical-chemical properties of the alkali chloride based systems / V. V. Rafalskiy. *Ukrainian Chemical Journal* – 1960. – V. 26. – pp. 585-587. (in Russain).

[20] Beliayev, A. N. Physical chemistry of molten salts / A. N. Belyayev, Ye. A. Zhemchuzhina. L. A. Firsanova. M., *Metallurguiya.* – 1957. – 312 p. (in Russian).

[21] Easteal, A. J. Electrical conductance of molten lead chloride and its mixtures with potassium chloride / A. J. Easteal, I. M. Hodge. *J. Phys. Chem.* – 1970. – №74 (4). – P. 730 – 735.

[22] Jinze, Li. Electrical Conductivity of LiCl–KCl–CsCl Melts / Li Jinze, Gao Bingliang, Chen Wenting, Liu Chengyuan, Shi Zhongning, Hu Xianwei and Wang Zhaowen. *Journal of Chemical Engineering Data.* – 2016. – 61 (4). – P. 1449-1453. doi: 10.1021/ САУ.jced.5b00682.

[23] Tarasova, N. M. Electrical conductivity of the $PbCl_2$-KCl system / N. M. Tarasova. *Journal of Physical Chemistry* – 1947. – V. 23. – 487p. (in Russian).

[24] Bloom, H. A note on the graphical analysis of survival data / H. Bloom, E. Heymann. *Proc. Roy. Soc. A.* – 1947. – №188. – C. 392 – 394.

[25] Lantratov, M. F. Electrical conductivity of molten salts. II. $PbCl_2$ – KCl system. / M. F. Lantratov, O. F. Moiseyeva. *Journal of Physical Chemistry* – 1960. – V. 34. – 367p. (in Russian).

[26] Apisarov, A. A. Electrical Conductivity of Molten Fluoride–Chloride Electrolytes Containing K_2SiF_6 and SiO_2 / Apisarov A. A., Redkin A. A., Zaikov Y. P., Chemezov O. V., Isakov A. V.. *Journal of Chemical Engineering Data.* – 2011. – 56. – P. 4733–4735.

[27] Korenko, Michal Electrical conductivity of systems based on Na3AlF6-SiO2 melt / Michal Korenko, Jozef Priscak, Frantisek Simko.

Chemical Papers. – 2013. – 67 (10). – P. 1350-1354. DOI: 10.2478/s11696-013-0393-x.

[28] Pavlovskii, V. A. Density and Electrical Conductivity of NaF-NaCl-WO$_3$ Melts / V. A. Pavlovskii. Inorganic Materials. – Vol. 39. – No. 11. – 2003. – P. 1208-1211. *Translated from Neorganicheskie Materialy.* – Vol. 39. – No. 11. – 2003. – P. 1394-1397.

[29] Kryukovsky, V. Electrical conductivity of low melting cryolite melts / V. Kryukovsky, A. Frolov, O. Tkacheva, A. Redkin, Yu. Zaikov, V. Khokhlov, A. Apisarov. *Light Metals.* - 2006. - P. 409.

[30] Danek, V. Density and Electrical conductivity of melts of the system Na$_3$AlF$_6$-AlF$_3$-LiF-Al$_2$O$_3$ / V. Danek, M. Chrenkova, A. Silny. *Proceedings the International Harald A. Oye Symposium.* Norway. - 1995. – P. 83.

[31] Dedukhin, A. Ye. Electrical conductivity of the [(KF-AlF$_3$)-NaF]-Al$_2$O$_3$ molten system / A. Ye. Dedukhin, A. P. Apisarov, O. Yu. Tkacheva, A. A. Redkin, Yu. P. Zaykov, A. V. Frolov, A. O. Gusev. *The melts.* – 2009. - № 2. - p. 18. (in Russian).

[32] Bakin, K. V. Electrical conductivity of the NaF-AlF$_3$-CaF$_3$-Al$_2$O$_3$ melts / K. B. Bakin, O. N. Simakova, P. V. Polyakov, Yu. G. Mikhalev, D. A. Simakov, A. O. Gusev. *the melts.* - 2009. - № 6. - p. 28. (in Russian).

[33] HUANG, You-guo Electrical conductivity of (N a$_3$AlF $_6$-40%K$_3$AlF $_6$)-AlF$_3$-Al$_2$O$_3$ melts / You-guo HUANG, Yan-qing LAI, Zhong-liang TIAN, Jie LI, Ye-xiang LIU, Qing-yu LI. *J. Cent. South Univ. Technol.* – 2008. – 15. – P. 819-823 DOI: 10.1007lsl 1771-008-0151-3.

[34] Yefremov, A. N. Properties of the electrolyte for black lead electrodes refining / A. N. Yefremov, A. P. Apisarov, P. A. Arkhipov, Yu. P. Zaykov. Studies in the field of refining and processing man-caused wastes: *proceedings of the All-Russian Conference with the Scientific School* (24-27 November, 2009), Ekaterinburg. – 2009. – pp. 85-88. (in Russian).

[35] Yefremov, A. N. *Electrical conductivity and liquidus temperature of the molten PbCl$_2$-KCl-PbO system* / A. N. Yefremov, A. P. Apisarov,

P. A. Arkhipov, Yu. P. Zaykov. *The melts.* – 2010. – № 1. – pp. 29-34. (in Russian).

[36] Redkin, A. Empirical evaluation and experimental investigation of chloride-oxide melts electrical conductivity of the system KCl-$PbCl_2$-PbO / A. Redkin, P. Arkhipov, A. Efremov, A. Apisarov. *Summaries 2 of the 19TH International congress of chemical and process engineering CHISA*, 28 August – 1 September 2010, Prague. – 2010. – P. 697.

[37] Yefremov, A. N. Electrical conductivity, density and liquidus temperature of the KCl-$PbCl_2$ equimolar mixture with lead oxide additions / A. N. Yefremov, P. A. Arkhipov. *Proceedings of the XIV Russian Conference "Structure and Properties of metal and slag melts,"* Ekaterinburg, IM UB RAS, 21-25 of September 2015. – 2015. – pp. 261-262. (in Russain).

[38] Yefremov, A. N. Electrical conductivity, density and liquidus temperature of the KCl-$PbCl_2$ equimolar mixture with lead oxide additions / A. N. Yefremov, N. P. Kulik, A. P. Apisarov, A. A. Redkin, A. yu. Chuikin, P. A. Arkhipov, Yu. P. Zaykov. News of Higher Educational Institutions. *Non-ferrous Metallurgy* – 2016. – №5. – pp. 10-16. (in Russian).

[39] Zabłocka-Malicka, M. Electrical conductivity of molten KCl–$DyCl_3$ system – comparison with other KCl–$LnCl_3$ systems / M. Zabłocka-Malicka, W. Szczepaniak, B. Ciechanowski. *Electrochim. Acta.* – 2013. – 114. – P. 424–429.

[40] Zabłocka-Malicka, M. Electrical conductivity of molten cesium chloride–dysprosium(III) chloride system / M. Zabłocka-Malicka, W. Szczepaniak. *J. Molecular Liquids.* – 2015. – 208. – P. 47–51.

[41] Janz, G. J. Thermodynamic and transport properties for molten salts: correlation equations for critically evaluated density, surface tension, electrical conductance, and viscosity data / G. J. Janz. *J. Phys. Chem. Ref. Data.* – 1988. – 17 (Suppl. 2). – P. 187 (ref. V. A. Khokholov, M. V. Sirnov, Zhur. Priklad. Khim. – 1970. – 43. – P. 302).

[42] Kovalevskii, A. V. Temperature dependence of electric conductivity of molten binary mixtures of alkali and rare-earth metals chlorides / A.

V. Kovalevskii, V. I. Shishalov. *Russ. J. Phys. Chem.* – 2006. – 80 (3). – P. 449–452.

[43] Potapov, A. M., L. Rycerz, E. S. Filatov, M. Gaune-Escard, Electrical conductivity of melts containing rare-earth halides. II. $MCl–PrCl_3$ (M = Li, Rb, Cs), *Z. Naturforsch.* 68a (2013) 59–65.

[44] Potapov, A. M. Electrical conductivity of melts containing rare-earth halides. I. $MCl–NdCl_3$ (M = Li, Na, K, Rb, Cs) / A. M. Potapov, L. Rycerz, M. Gaune-Escard. *Z. Naturforsch.* 2007. – 62a. – P. 431–440.

[45] Antipin, L. N. *Electrochemistry of molten salts* / L. N. Antipin, S. F. Vazhenin. – M.: State scientific technological publishing house for black and non-ferrous metallurgy literature, 1964. – 356 p. (in Russian).

In: An Essential Guide …
Editor: Luke Lewin

ISBN: 978-1-53615-047-6
© 2019 Nova Science Publishers, Inc.

Chapter 2

SUPERCAPACITOR ELECTRODES OF ACTIVATED CARBONS FROM NATURAL SOURCES

Md. Shahnewaz Sabit Faisal and Ramazan Asmatulu*

Department of Mechanical Engineering
Wichita State University, Wichita, KS, US

ABSTRACT

Activated carbon (AC) is carbon that is processed at elevated temperatures with activating agents. It has relatively low-volume pores, meaning increased surface area, in order to store a considerable amount of energy. In addition to coal and coke, AC can also be produced from natural materials, which can be inexpensive and sustainable natural sources for utilization in energy storage devices, such as supercapacitors and battery anodes. Many nutshells, coconut shell, bamboo, sugarcane bagasse, rice bran, corn cob, and potato wastes can be good sources of AC for different applications. These natural materials are used as a precursor, carbonized at high temperature in an oxygen free environment, and activated by physical or chemical activation processes to produce AC with a very high specific

---

* Corresponding Author's E-mail: ramazan.asmatulu@wichita.edu.

surface area ($\sim$1500 m^2/g). The activated carbon is used to fabricate electrodes of supercapacitors and is characterized using different techniques prior to the application. These supercapacitors have a high specific capacitance, high energy density, and high power density compared to many other storage devices. This book chapter reports on recent developments in stabilization, carbonization, and activation of naturally grown biomass; their physical and chemical properties; and major applications in supercapacitors. This study shows that natural materials are promising options for producing ACs for different supercapacitor applications at a low cost. Many readers, such as students, engineers, scientists, and other participants in supercapacitor technologies, will greatly benefit from this work.

Keywords: natural materials, activated carbon, specific surface area, specific capacitance supercapacitor

1. INTRODUCTION

With the advancement of technology, our dependency on energy storage devices has been steadily increasing. There are several ways to store electrical energy, including capacitors, supercapacitors, batteries, and fuel cells, which have been extensively used in various industrial applications [1, 2]. Supercapacitors have the potential to be a major advancement in energy storage and can fill the gap between conventional capacitors and rechargeable batteries. The use of supercapacitors, also referred to as electric double-layer capacitors (EDLCs) and ultracapacitors, is a promising option for charging and discharging energy rapidly whenever it is needed [1-6].

The type of electric energy storage typically requires an evaluation based on the application, life span, cost, performance, and power density in terms of charging and discharging. In comparison to regular high-power lithium batteries, supercapacitors are able to deliver more specific power and higher specific energy. They also show very high life cycles and a wide range of operating temperatures compared to regular batteries [4]. Supercapacitors can charge and discharge so rapidly that they can provide a backup source of high peak power in seconds. Because they have longer life cycles and are lighter and functional, they do not require maintenance or replacement, as is the case with batteries. These devices are more environmentally friendly because there is no use of heavy metals and no toxic disposal issues. Because of these unique characteristics and with the increase in its energy density, supercapacitors are being viewed as an attractive alternative to regular batteries [5]. Figure 1 shows the commonly known Ragone plot of different energy-storing devices [1]. Conventional capacitors show higher power density but lower energy density, and batteries show a lower power density but higher energy density. Supercapacitors fall in between these two choices of electric energy-storing devices.

Hermann von Helmholtz, a German physicist, first described the double-layer capacitor in 1853. In 1957, the General Electric Company first patented an electrochemical capacitor (U.S. Patent 2800616), which had a double-layer capacitance and porous carbon for the electrodes [7-9]. In 1961, the Standard Oil of Ohio (SOHIO) Research Center (Cleveland, USA) started working on the commercial double-layer capacitor. SOHIO continued until 1971 when they developed disk-shaped capacitors. Due to the lack of marketing and interest, they licensed their technology to Nippon Electric Company (NEC) of Japan, who first introduced their electrochemical products as memory-backup devices [7]. In 1978, Panasonic made a commercially successful electrochemical capacitor named "Goldcaps." The next year, Elna Co. Ltd., introduced first-generation EDLCs called "Dynacaps," which had high internal resistance. These were used for data backup or low-current applications. The Pinnacle Research Institute (PRI) developed the first supercapacitor, called the "PRI Ultracapacitor," with low internal resistance in 1982 for use in military applications. Later in 1992, Maxwell Laboratories took over the development of the PRI Ultracapacitor for power applications. Apart from these companies, currently a few others around the world are manufacturing EDLCs. In the U.S., Kold Band International, Epcos, ELNA, and AVX are commercially manufacturing EDLCs [10-14].

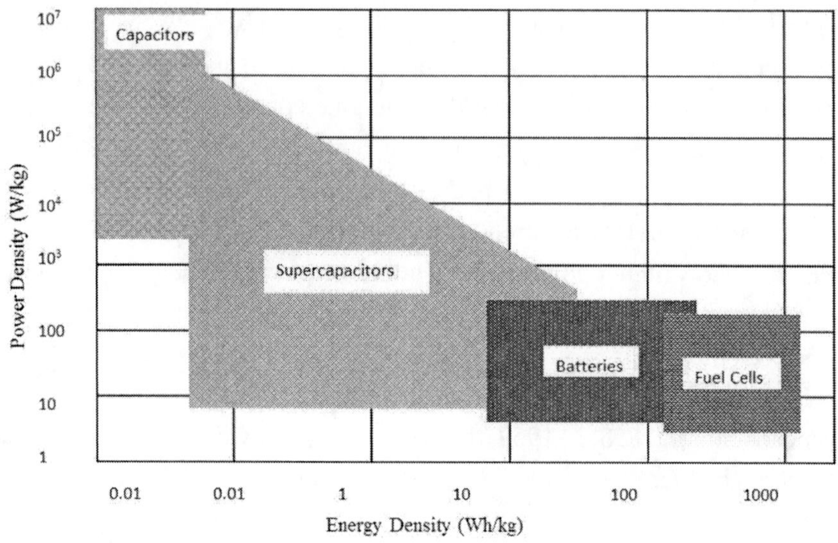

Figure 1. Ragone plot of energy storage systems, adapted and redrawn from reference [1].

Compared to other conventional energy-storing devices, the supercapacitor electrode has an extremely large surface area for storing electric energy. This has opened a wide range of options for using different materials having a high specific surface area, such as activated carbons (ACs). Using natural sources to produce ACs is also environmental friendly and low in cost. Supercapacitors work on the same principle as conventional electrostatic capacitors. A conventional capacitor consists of two conductors separated by a non-conductive region called the dielectric. When electrical energy is applied across the conductors, energy is stored as a form of electrical charge and later acts as a source of energy when needed. In supercapacitors, the surface area of the conductors is much higher than in conventional capacitors. The capability of a device for charge storage per unit voltage is known as capacitance. If voltage created across the conductors is V and charge is stored in conductors $\pm Q$, then capacitance C would be

$$C = \frac{Q}{V} = \varepsilon \frac{A}{d} \tag{1}$$

where A is the conductors' surface area, d is the distance between conductors, and ε is the dielectric constant of the dielectric material [6, 7, 16]. Also, the stored energy (E) in a capacitor can be calculated as

$$E = \frac{1}{2}CV^2 \tag{2}$$

Electrical energy can be stored in two different ways: indirectly in batteries, through Faradic oxidation and reduction, which supplies energy throughout the electrochemical process; and also directly, by storing negative and positive charges in capacitors/supercapacitors, also known as non-Faradic electric energy storage [1]. An important difference between capacitors and batteries is the reversibility. In batteries, the electric energy is produced from Faradic electrochemical reactions through phase changes of chemical substances. Although the overall thermodynamic energy conversion can be conducted in a relatively reversible way, the charge and discharge processes often involve irreversibility in the interconversion of chemical reagents, which results in a restricted cycle life. Depending on the type of battery, cycles can be one thousand to several thousand. On the other hand, capacitors/supercapacitors have almost unlimited life cycles because chemical reactions or phase changes are not involved [6]. From Figure 1, it is clear that supercapacitors have higher power densities than batteries. It should be noted that there has never been an aim to replace batteries with supercapacitors, but electrically coupling electrochemical capacitors in discharge and recharge with batteries offer complementary opportunities [6]. Supercapacitors are divided into two groups depending on the mechanism for storing charge. One group is pseudocapacitors, and the other group is electric double-layer capacitors.

The pseudocapacitor is based on the working principle of a battery. Electrical energy is obtained by progressive redox reactions between several oxidation states. Charge and discharge occurs not only on the surface of but also throughout the electrodes. This capacitor is also called a Faradic supercapacitor [15-18].

The electric double-layer capacitor has two electrodes impregnated with an electrolyte. Its working principle is different from that of a conventional battery in that the EDLC directly stores electrical charges by physically separating the positive and negative charges on the electrode surface. When electric potential is applied on the electrodes, electrons from the negative electrode move through the electrolyte to the positive electrode, and positive ions from the positive electrode move to the negative electrode. The charges are held statically on the surface of the electrodes. Unlike the pseudocapacitor, the EDLC holds charges only on its surface, not throughout the electrodes. Because the charge-and-discharge technique only involves the physical separation of ions, no chemical reaction occurs in the EDLC, which eventually leads to an almost unlimited cycle life without any failures [6, 17, 18].

2. SUPERCAPACITOR MODELS

Various supercapacitor models have been developed to explain the electrical process that occurs at the boundary of the solid conductor and the electrolyte. Three popular models that explain that process are discussed in this section.

2.1. Helmholtz Model

Helmholtz first introduced the "double-layer" model to explain the electrode/electrolyte interface in capacitors. Figure 2 illustrates the schematic view of the Helmholtz double-layer model that shows the locations of positive and negative ions on the surfaces [7].

In the Helmholtz double-layer model, the electrode acquires a layer of positive charge on its surface, and next to that surface is a layer of electrons in the electrolyte. The same interface occurs at the other electrode but with opposite charges to create a charge balance. Because of this two-layer formation in the capacitors, Helmholtz called this as a double-layer behavior. The double layer can be thought of as a molecular capacitor [7,

14, 18-22]. According to the Helmholtz model, the differential capacitance (C_l) is given by [14]

$$C_l = \frac{\varepsilon}{4\pi\delta} \tag{3}$$

which implies that the differential capacitance C_n is a function of dielectric constant and charge layer separation (δ).

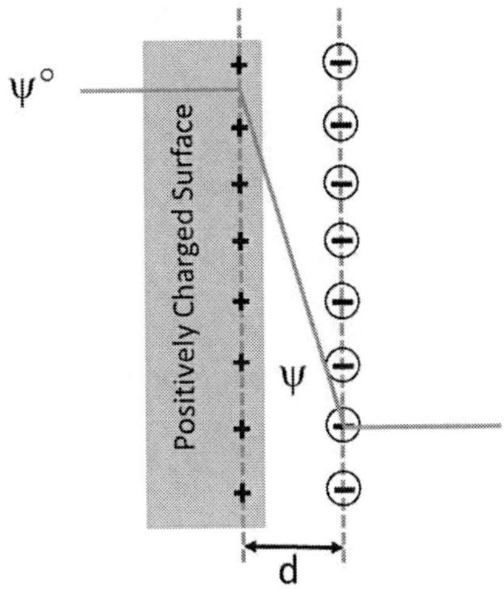

Figure 2. Schematic illustration of Helmholtz double-layer model, adapted from [7].

2.2. Gouy-Chapman Model

The Gouy-Chapman model proposes that capacitance is not constant but rather dependent on applied potential and ionic concentration. This model maintains that thermal motion creates a space distribution of ions in electrolytes. Both coulombic force and thermal motions influence the equilibrium distribution of counter ions. The counter ion concentration decreases, while a similar ion concentration increases from the electrode surface. This phenomenon is referred to as the diffuse electric layer [7, 14, 22]. Figure 3 shows a schematic view of the Gouy-Chapman diffuse model.

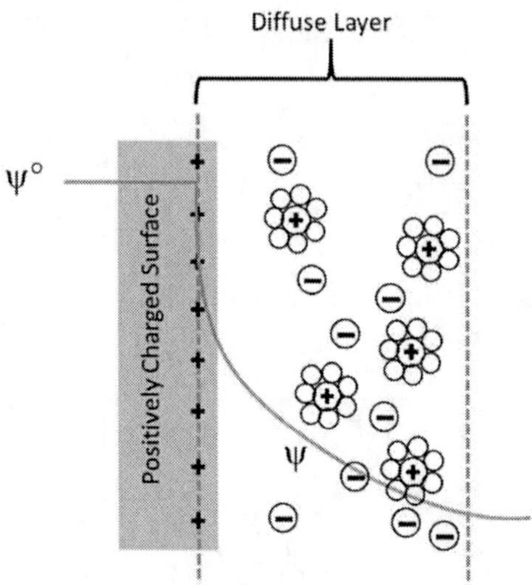

Figure 3. Schematic view of Gouy-Chapman diffuse model, adapted from [7].

This model was developed by combining Poisson and Boltzmann equations. According to this theory, differential capacitance (C_G) can be explained by equation (4):

$$C_G = \frac{\varepsilon\,\kappa}{4\,\pi}\cosh\frac{z}{2} \tag{4}$$

where z is the valence of ions, and κ is the reciprocal Debye-Hückel length and can be defined by

$$\kappa = \sqrt{\frac{8\pi n e^2 z^2}{\varepsilon k T}} \tag{5}$$

where T is the absolute temperature, n is the number of ions per cubic centimeter, and k is Boltzmann's constant [14]. This double-layer model is not as compact as the Helmholtz model, because ions are considered as highly mobile point ions, and metal is a perfect conductor for that reason. This model leads to an overestimation of electric double-layer capacitance.

2.3. Stern Model

In 1923, O. Z. Stern combined the Helmholtz model and Gouy-Chapman model by introducing two regions of ion distribution in a graphical form. Figure 4 shows the schematic illustration of the Stern model.

The inner region is a compact layer, or Stern layer, and the outer region is called the diffuse layer, or Gouy-Chapman layer. The total capacitance (C) of this EDLC can be calculated by combining the capacitance of these two regions [7, 22] as

$$\frac{1}{C} = \frac{1}{C_l} + \frac{1}{C_G} \tag{6}$$

where C_l is the Stern layer capacitance, and C_G is the Gouy-Chapman layer capacitance [14]. In a practical EDLC, the electric double-layer behavior is much more complex in porous electrodes. This behavior can be affected by a number of factors, such as space constraints inside the pores, ohmic resistance of the electrolyte, wetting behavior of the porous electrode, and the mass transfer path [7].

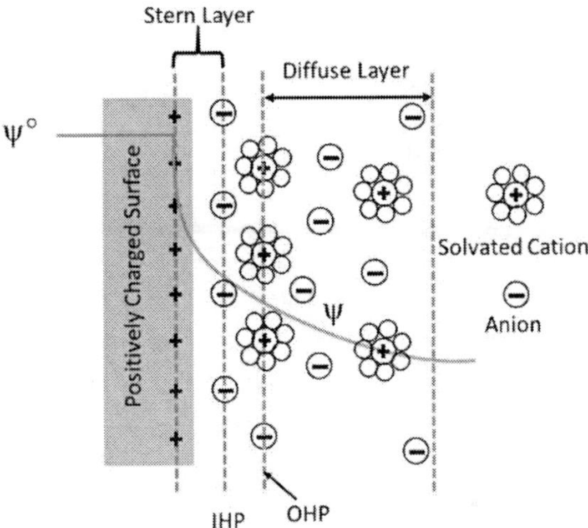

Figure 4. Schematic illustration of the Stern model, adapted from [7].

2.4. Current Model

In 1963, Bockris, Devanathan, and Muller proposed a double-layer structure that considered the action of solvent. They proposed that specific adsorption occurs at the inner Helmholtz plane, hydrated anions and cations occupy the Gouy plane, and the occupancy of any ions in the inner Helmholtz and Gouy planes is prohibited. The surface potential decreases in both the inner Helmholtz and Gouy planes and the diffuse double layer [14, 20, 22]. Figure 5 depicts a schematic view of the double-layer model including layers of solvent.

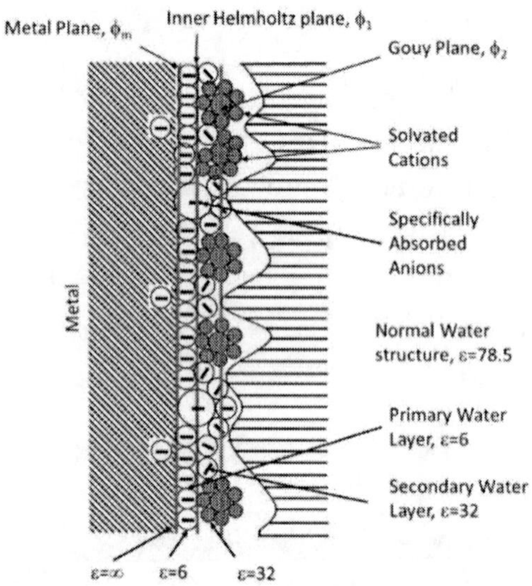

Figure 5. Schematic illustration of double-layer model showing layers of solvent, adapted from [20].

3. ELECTRIC DOUBLE-LAYER CAPACITORS

3.1. Construction of EDLC

A practical double-layer capacitor is comprised of two electrodes connected to a current collector and separated by a separator soaked in electrolyte. A wide variety of materials is used for constructing an EDLC. The basic construction of a carbon-based EDLC and its actual model are shown in Figure 6 [7]. It uses porous paper as a separator, aluminum foil as a current collector, and a PET sheet to insulate the device from any other media outside.

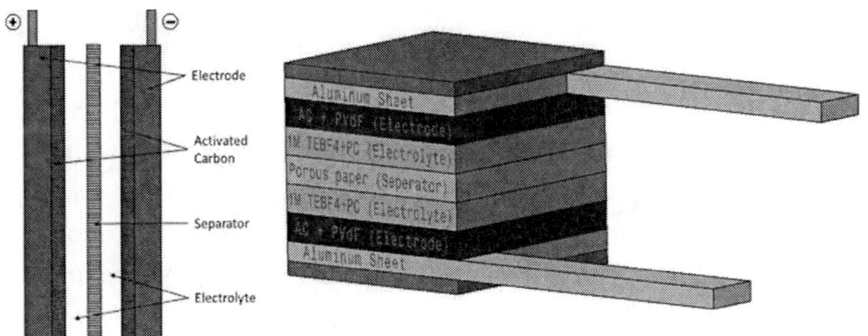

Figure 6. Diagrams showing (a) basic construction of EDLC, adapted from [7], and (b) actual model of EDLC.

3.2. Electrode Materials

3.2.1. Key Materials and Properties

The key material to a high-performing EDLC is in its electrode material. The higher the surface area of the electrodes, the more charge they can contain for better energy storage. The selection of electrode material is very important. Supercapacitors are categorized based on the materials used for their electrodes: carbon, metal oxides, and conducting polymers. Since carbon materials can meet the requirements as electrode materials for an EDLC, various carbonaceous materials such as activated carbons, aerogels, xerogels, nanofibers, nanotubes, and graphite have been studied extensively [7].

Carbon is the most popular electrode material used for supercapacitors. Typically it is used with various modifications. Activated carbon is the most commonly used type of porous carbon utilized for supercapacitors because of its high surface area, low cost, easy production in larger quantities, great cycle stability, availability, wide operating temperatures, non-toxicity, and established electrode production technologies. AC in powders, woven cloths, felts, or fibers can have a surface area up 1500–2500 m²/g. The charge storage in AC is mostly capacitive in the EDLC. Also, surface functional groups in ACs that can charge and discharge give rise to pseudocapacitance [1, 7]. Coal, pitch, and petroleum coke are the most commonly used materials for the commercial production of AC. Nonetheless, the scarcity of fossil fuels, environmental pollution issues, and cost have led AC productions to sustainable and renewable resources. Therefore, wood, nutshell, cellulose, coconut shell, bamboo charcoal, bamboo fibers, corn grain, rotten potato, rice bran, banana fiber, and sugar cane bagasse are popular sources used for producing AC. The cost of some of these raw materials and carbon yields from pyrolysis of these sources are presented in Table 1 [23].

Theoretically, the higher the specific surface area of activated carbons, the higher the expected specific capacitance, but in reality, this situation is more complicated. One type of AC with a small surface area can give a larger specific capacitance compared to the type with a larger surface area. This is because actual double-layer capacitance varies depending on the process used for activating the carbons. The porous texture of the carbon determines the conductivity of ions in a capacitor. The mobility of ions can be affected in small pores when compared to large pores. Also, the ion mobility is different in the bulk of the electrolyte. The pore size must be chosen to suit the electrolyte and to ensure that it is optimal, depending on the size of the ions [24-28].

Electrochemical capacitors made from carbon are of two types, based on the kind of energy accumulated. One type is the electrical double-layer capacitor where only the electrostatic attraction between ions and the charged surface occurs, and the other type is a supercapacitor based on additional Faradic pseudocapacitance reactions. The performance of a supercapacitor simultaneously combines these two kinds of energy. Figure 7 illustrates the typical charge/discharge voltammetry of an electrochemical capacitor. An ideal double-layer capacitor is in the form of a rectangle, where the sign of the current is immediately reversed after reversing the voltage sweep. However, electrode materials with pseudocapacitance characteristics deviate from this rectangular shape, showing peaks caused by redox reactions [28].

Table 1. Raw material cost and carbon yields of pyrolysis of some natural precursors [23]

Raw Material	Raw Material Cost ($/kg)	Carbon Yield from Pyrolysis (wt%)
Petroleum Coke	1.4	90
Charcoal	1.2	90
Lignite	0.75	50
Coconut Shell	0.25	30
Wood	0.8	25
Potato Starch	1.0	45
Sucrose	0.25	<45
Cellulose	0.65	<45
Corn Grain	0.25	<45
Banana Fiber	4	<45

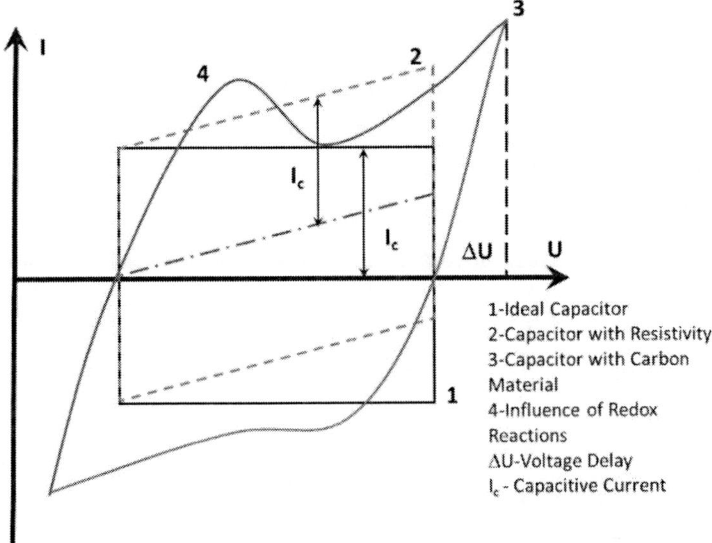

Figure 7. Typical charge/discharge voltammetry of electrochemical capacitor [28].

3.2.2. Bamboo

Bamboo biomasses can be used to make activated carbon. Clean bamboo charcoal is ground to make fine powder. Then, HCl acid is used to remove any organic residue. Yang et al. [29] used 3M HCl to soak the bamboo charcoal powder for 12 hours. The powder is washed with DI water to eliminate any acid and dried overnight. It is then heated at high temperature to carbonize the powder in an inert environment at a constant heating rate. Usually this temperature ranges from 600–1000°C. The carbonized powders are mixed with KOH and heated around 400°C for one hour to expand the surface area by activation. X-ray diffraction (XRD) is used to investigate the crystal structure. In the XRD analysis, carbon particles without KOH activation show a graphitic structure, but KOH-activated carbon does not show a graphitic structure. The activation creates a porous structure with smaller fragments, breaks the [002] and [100/101] planes, and introduces some amorphous carbon.

Surface texture can be studied with high magnification using scanning electron microscopy (SEM). A non-activated bamboo charcoal carbonized at 900°C shows relatively flat and smooth surfaces, whereas an activated carbon shows a sponge-like structure with wrinkles and pores. The surface area, pore volume, and average pore size is determined using the Brunauer-Emmett-Teller (BET) technique. Non-activated carbon carbonized at 900°C shows a BET surface area of 70 m^2/g, and activated carbon with the same parameters shows a BET surface area of 2960 m^2/g. As is stated, KOH activation substantially increases the surface area. The increased surface area of activated carbon provides a vital role while it is used as electrodes in supercapacitors. It shows a prominent increase in specific capacitance than the non-activated carbon. This activated carbon carbonized at 900°C indicates a high specific capacitance of about 300 F/g at 0.1 A/g current density in an aqueous electrolyte. Galvanostatic charge-discharge measurement shows 92% capacitance retention at 2 A/g current density after 3000 cycles, which provides excellent cycle stability for energy storage devices. Yang et al. suggested that common bamboo can be an essential raw material for future energy storage applications [29].

Bamboo fiber is activated with KOH (1:1 ratio) and carbonized in two steps. Final carbonization occurs at 800°C for one hour in an inert environment in the presence of Argon. These carbonized bamboo fibers resemble the graphitic phase in an XRD pattern analysis. D and G bands of carbon are found in further analysis with Raman spectroscopy. A higher G band intensity compared to D band intensity signifies that the carbonized fibers are rich in the graphitic phase. The graphitic phase is significant because it is a conducting form of carbon, whereas the diamond phase is a non-conducting form of carbon. The BET technique indicates a surface area of 1120 m^2g^{-1} with 0.34 cm^3g^{-1} pore volume and 1.22 nm pore size. SEM image analysis shows that it has a highly porous surface with 10 μm average diameter. The porous structure could originate with the KOH activation following this equation [30, 31]:

$$6KOH + C \rightarrow 2K + 3H_2 + 2K_2CO_3$$

A further study of the surface texture with transmission electron microscopy (TEM) shows an interconnected, three-dimensional random morphology [30]. Specific capacitance (C_{sp}) can be investigated using the Galvanostatic charge-discharge technique and is calculated from the following equation:

$$C_{sp} = \frac{I\Delta t}{m\Delta V} \tag{7}$$

where I in amps, Δt in seconds, m in grams, and ΔV in voltage are the discharge current, discharge time, mass of the active materials, and potential windows, respectively [32]. In this chemical characterization, KOH, NaOH, and LiOH are used as electrolytes. In the Galvanostatic charge-discharge technique at a current density of 0.4 A/g, the specific capacitance for KOH, NaOH, and LiOH are 512, 202, and 156 Fg^{-1}, respectively. The ionic radii decrease from K^+ to Li^+ ($K^+ > Na^+ > Li^+$), but in an aqueous medium, Li^+ has the largest radius. It is believed that Li^+ has less mobility compared to K^+ and Na^+; as a result, it gives the lowest specific capacitance. These carbonized bamboo fibers show a high power density of 7.9 kW/kg and energy density of 54 kWh/g [30]. The energy density (E) and power density (P) are expressed as follows [33]:

$$E\left(\frac{Wh}{kg}\right) = \frac{1}{2}\,C_{sp} \times \Delta V^2 \times \frac{1000}{3600} \qquad (8)$$

$$P\left(\frac{W}{kg}\right) = \frac{E \times 3600}{t} \qquad (9)$$

where C_{sp}(F/g) is the specific capacitance measured from the Galvanostatic charge-discharge technique, ΔV(V) is the potential window, and Δt(s) is the discharge time. These fibers show a very stable performance of about 100% over 5,000 charge-discharge cycles. The supercapacitor shows no degradation in charge storage capacity on bending and indicates an improved performance at higher temperature. It is observed that performance improves about 65% while increasing the environmental temperature from 10 to 70°C. The activated bamboo fibers are potential material for energy storage devices that can operate over a wide operating temperature [30-33].

Bamboo-based carbons can also be activated at 900°C for one hour with steam and non-aqueous electrolyte solutions. This provides a high surface area and specific capacitance from 445 to 1,025 m²/g and from 5 to 60 F/g, respectively [34].

3.2.3. Coconut Shell

Coconut shell is used as a fuel and is also a good source of charcoal. Activated carbon is produced from coconut shell because of its mesoporous structure, which is suitable for supercapacitor applications. Crushed coconut shell powder is dried at 300°C for 5 hours, then soaked in $CaCl_2$ (25 wt%) solution for 18–20 hours to produce activated charcoal. It is chemically activated with KOH solution at 600°C to produce activated carbon. The structural analysis of this activated carbon shows a high specific surface area. It also shows a 1640 m²/g BET surface area, 0.7522 cm³/g micropore volume, and 3 nm pore size. XRD analysis indicates that this activated carbon is rich in the graphitic phase and amorphous carbon. Electrochemical studies also indicate that it has a specific capacitance of 534 mF/cm^2, which is equivalent to a single-electrode specific capacitance of 356.2 F/g. It has high energy and power density of 88.8 Wh/kg and 1.63 kW/kg, respectively. It shows stable capacitance up to about 600 cycles, which is relatively low compared to bamboo-based activated carbon [35].

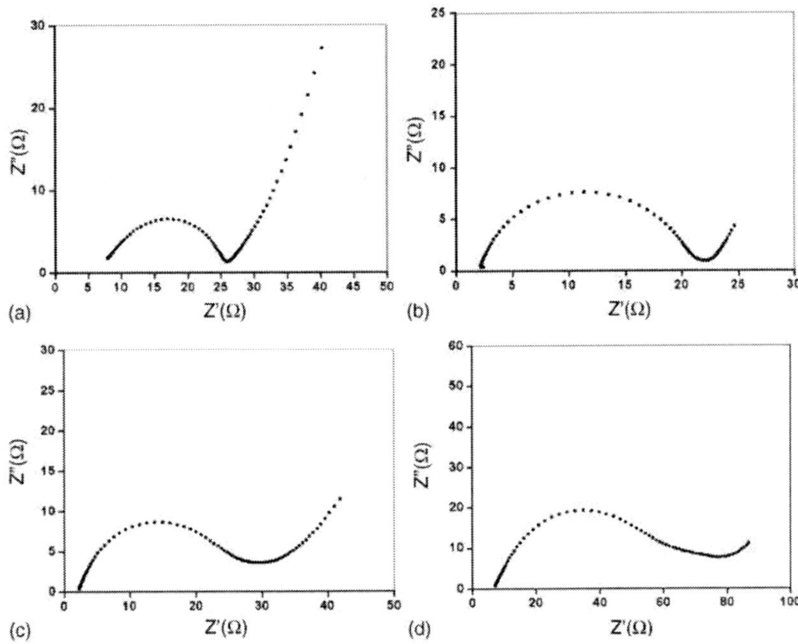

Figure 8. Nyquist plots: (a) activated carbon compared with (b) 1.6 wt %, (c) 5 wt %, and (d) 7 wt % of RuO$_2$·xH$_2$O [36].

The specific capacitance of this activated carbon can also be enhanced by incorporating hydrous ruthenium oxide (RuO$_x$(OH)$_y$) in various amounts and then homogenously dispersed into the porous structure of activated carbon. The sample (75 wt%), graphite (20 wt%), and ethyl cellulose as binder (5 wt%) are mixed together to fabricate electrodes for supercapacitors. XRD patterns show that the hydrous ruthenium oxide is in an amorphous state in the composite, which significantly increases the specific capacitance. In electrochemical analysis, the composite electrode in

1M H_2SO_4 is used as a working electrode, $Hg|Hg_2SO_4$ as the reference electrode, and platinum foil as the counter electrode. There is a linear relationship between the specific capacitance and amount of ruthenium incorporated. Hydrous ruthenium oxide with 9 wt% shows a significant increase in the specific capacitance from 100 to 250 F/g. A Nyquist plot of activated carbon, illustrated in Figure 8, indicates some interesting behavior. Composite electrodes show a larger semicircle than that of the activated carbon electrodes. The diameter of the semicircle increases with the increment of $RuO_x(OH)_y$, which signifies that the increment of resistance is imposed by $RuO_x(OH)_y$. The straight line in the lower frequency region starts deviating from being a straight incline, which signifies that pores in the activated carbon are occupied by $RuO_x(OH)_y$. Although this increases the resistance considerably, the amorphous form of $RuO_x(OH)_y$ significantly increases the specific capacitance as well [36].

3.2.4. Pistachio Shell

Pistachio shells are natural waste material and an excellent precursor for producing activated carbon for supercapacitor applications. In recent studies, pistachio shells are washed several times, soaked in deionized (DI) water overnight, and then dried in an oven for two hours at 105°C to evaporate all water content [37]. Following this procedure, they are stabilized in an oven at 240°C at 5°C/min heating rate for two hours to evaporate any volatile content in the shells. Then the stabilized nutshells are immersed in a 1 wt% KOH solution, and the soaked shells are placed in a vacuum desiccator at 22 in Hg vacuum for 30 minutes to evacuate any trapped air. The nutshells are then soaked in the KOH solution for twelve hours for chemical activation again. Following this, the solution is drained out, and the nutshells are dried in an oven at 85°C for two hours prior to carbonization, which is carried out in a tube furnace, holding it at 700°C for two hours at 5°C/min heating rate with a continuous flow of argon gas. The flow of argon gas flow ensures that the carbonization process occurs in an inert environment. The carbonization process is also done at 900°C separately. The carbonized nutshells are washed at least three times with a 36% HCl solution to neutralize any KOH content in the shells, and then the shells are washed again at least three times with DI water to remove any additional HCl content. Subsequently, the shells are dried in an oven at 105°C for two hours to evaporate the water content and collect the activated carbons, which are ground using a mortar and pestle for thirty minutes to make activated carbon powders [37, 38].

During the supercapacitor fabrications, layouts (2x2 cm) 0.13 mm thick are created using Scotch tape on aluminum foil. Polyvinylidene fluoride (PVdF) is used as a binder with the activated carbon powders. The N, N-dimethylacetamide (DMAC) solvent and acetone (70:30 weight ratio) are mixed with 5 wt% PVdF and stirred with a magnetic stirrer for four hours at 50°C on a hot plate. Then the AC is dispersed into a PVdF solution (50:50 weight ratio) using a magnetic stirrer for five minutes. The slurry is placed on aluminum foil layouts, maintaining the dimensions, to construct the carbon electrodes. Finally, the electrodes are kept in oven at 120°C for four hours to dry the slurry mixture and make the solid carbon electrodes. The organic electrolyte is prepared using 1M ammonium salt, called tetraethylammonium tetrafluoroborate (TEABF$_4$), and propylene carbonate (PC). Lens tissues (2x2 cm) are immersed into the electrolyte and kept in a vacuum desiccator to properly soak for 30 minutes. The lens tissues are used to contain the electrolytes. The EDLC is constructed following the model shown previously in Figure 6(b).

Carbonization of pistachio shells at 700°C and 900°C shows that there is no significance difference in carbon yield. Both temperatures yield 60–61% carbon, indicating that carbonizing can be followed using 700°C rather than 900°C, in order to be more economical. Analysis of the surface texture of the carbons confirms that KOH activation produced a highly porous surface area, the most important factor for carbon electrodes in supercapacitors. SEM is used to inspect the surface texture of the activated pistachio. Non-activated carbonized shells do not show any porous surface texture; only a lamella surface is created. Activated carbons show highly porous surfaces for both carbonization temperatures. Activated pistachio shell carbonized at 700°C shows a more porous surface area than that carbonized at 900°C. Shell carbonized at 900°C shows larger pores than that carbonized at 700°C. It is believed that the high-temperature (900°C) carbonization might melt more of the carbonaceous materials of the shells and create larger pores with a stiffer structure. Figure 9 illustrates the non-activated and activated pistachio shell in SEM analysis under 10 μm magnification.

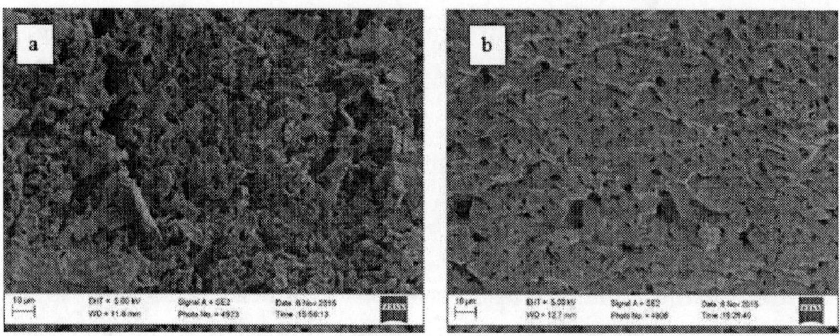

Figure 9: SEM images of pistachio shell: (a) non-activated pistachio shell carbonized at 700°C and (b) cross-section of activated pistachio shell carbonized at 900°C.

In the capacitance technique at a scan rate of 10 mVs^{-1}, activated carbon carbonized at 700°C shows a specific capacitance of 27,083 μF/g and at 900°C shows this to be 14,583 μF/g. In a Nyquist plot, AC carbonized at 900°C shows a larger semicircle than that carbonized at 700°C. It is observed that activated carbon carbonized at 700°C performs better for this purpose, compared to the one carbonized at 900°C, and also uses less energy in the carbonization process. It is found that pistachio shell is a potential source of producing activated carbon for supercapacitor applications [37, 38].

3.2.5. Acorn Shell

Acorn shell is natural waste and an excellent precursor in producing activated carbon. It is carbonized at two different high temperatures of 700°C and 900°C to study the carbonization temperature effect on electrochemical behavior. The process and parameters of making the activated carbon and EDLC are identical to that used for pistachio shell. Carbonization of acorn shells at 700°C yields 76% carbon and at 900°C yields 80% carbon. In the SEM image analysis, the non-activated carbonized acorn shell does not show any porous surface texture, but activated carbon shows a highly porous surface for both carbonization temperatures. In a comparative study, AC carbonized at 700°C shows more porous surfaces than that carbonized at 900°C. In the CV technique at a scan rate of 10 mVs^{-1}, AC carbonized at 700°C shows a specific capacitance of 50,000 µF/g and that carbonized at 900°C shows 16,666 µF/g. The same phenomenon as for pistachio shell were observed. It is found that activated carbon carbonized at 700°C performs better for this purpose than AC carbonized at 900°C. Acorn shell is another potential source of producing activated carbon for supercapacitor applications. Figure 10 shows SEM images of non-activated and activated acorn shells [37, 38].

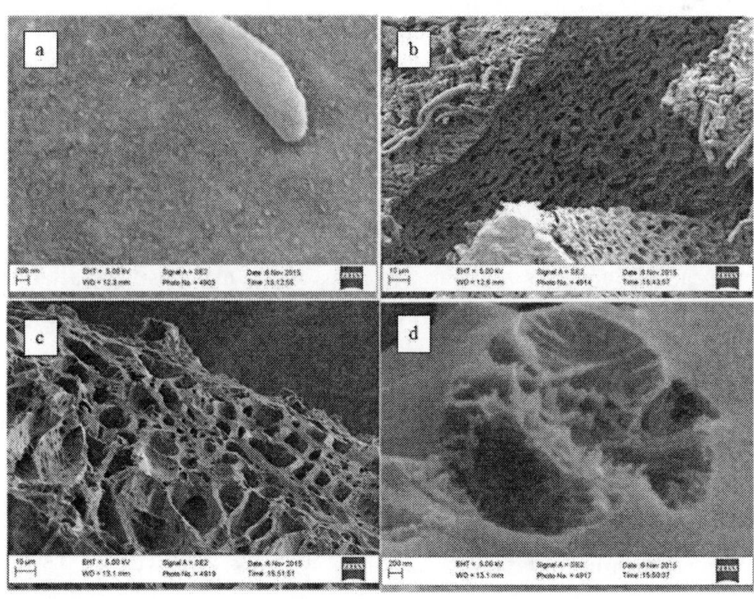

Figure 10. SEM images of acorn shell: (a) non-activated carbonized at 900°C at 200 nm magnification, (b) cross sections carbonized at 700°C with 10 µm magnification, (c) cross sections carbonized at 900°C with 10 µm magnification, and (d) cross sections carbonized at 900°C at 200 nm magnification.

3.2.6. Rice Bran

The world has an abundance of rice bran, an agricultural by-product and good source of producing activated carbon. Rice bran is carbonized at 700°C for one hour with the presence of N_2 gas and then activated with KOH in the same environment at 850°C for one hour. SEM and TEM analyses show that the activation process makes a smooth carbonized rice bran to an interconnected honey-comb like structure with meso/micropores. Researchers studied different mass ratios of solid KOH to carbonized rice bran to find a suitable combination. It is found that maximum BET surface area of 2,475 m^2/g and pore volume 1.21 cm^3/g is achieved with a KOH to carbonized rice bran ratio of 4. This combination exhibits a well-balanced micro/mesopore interconnected structure. The electrodes are prepared using 85 wt% of the activated carbon, 10 wt% of acetylene black, and 5 wt% of polytetrafluoroethylene (PTFE) as a binder. A quantity of 6M KOH is used as an aqueous electrolyte, and 1-Ethyl-3-methylimidazolium tetrafluoroborate (EMIMBF4) is used as an ionic electrolyte. The aqueous electrolyte shows a high specific capacitance of 265 F/g and 182 F/g with high current densities of 10 A/g and 100 A/g, respectively. The ionic liquid shows a stable lifecycle over 10,000 cycles, specific energy of 70 Wh/kg, and specific power of 1,223 W/kg [39].

3.2.7. Sugarcane Bagasse

Sugarcane bagasse is a cheap biomass waste and easily available. It is good source for making activated carbon. Researchers produce activated carbon from sugarcane bagasse using facile microwave-induced $ZnCl_2$ activation. This activated carbon is studied by varying the percentage of $ZnCl_2$ to examine its surface texture and electrochemical behavior for supercapacitor applications. Bagasse pith is washed in DI water and dried at 100°C overnight. In an N_2 atmosphere, the precursor is loaded with $ZnCl_2$ and subjected to microwave irradiation (2.45 GHz, 700 W) for 15 minutes to produce activated carbon. PTFE (5 wt%), as a binder, is mixed with activated carbon to fabricate electrodes. Researchers used EMIMBF4 as an electrolyte and varied $ZnCl_2$ from 20 wt% to 60 wt% for activation. The SEM analysis shows small mesopores observed with 20 wt% of $ZnCl_2$, but as this percentage increases, large mesopores and macropores are indicated. $ZnCl_2$ is responsible for degradation of cellulose material and aromatization of the carbon skeleton [40, 41]. The CV technique shows a higher specific capacitance for 60 wt% $ZnCl_2$ than its lower weight percentages. This may be due to the large ions and ion-pairs of EMIMBF4 being able to move easily. This is also verified with capacitance retention, which is used to evaluate the specific ion transfer behavior: the larger the retention ratio, the better the ion transfer behavior. A 60 wt% concentration of $ZnCl_2$ showed the highest capacitance retention and 20 wt% $ZnCl_2$ showed the lowest capacitance retention at any frequency. In a Nyquist plot, the semicircle size decreases as the weight percentage of $ZnCl_2$ decreases from 60 wt% to 20wt%. This signifies that the large pore size favors the diffusion of ions. Capacitive performance of nanoporous carbons is intimately related to pore size. Pore sizes can be easily tuned in a wide range by adjusting the weight percentage of $ZnCl_2$ during activation [40].

3.2.8. Potato Waste

Rotten potato waste is used as precursor to prepare three-dimensional activated carbon. KOH is used for chemical activation followed by hydrothermal treatment and carbonization. This activated carbon shows superior cycling stability and high capacitance of 269 F/g at 1 A/g current density. Also, the assembled symmetric device gives a high specific capacitance of 48 F/g and high energy density of 4.27 Wh/kg at 1 A/g current density. This facile and low-cost method is a potential for commercial supercapacitor applications [41, 42]. Apart from these natural sources, activated carbons are also produced from areca fibers and activated with KOH and steam to use as electrode material for supercapacitors [27]. In addition, a number of other natural materials (e.g., algae, peanut shell, fruit peels, etc.) have been considered for supercapacitor and energy storage applications [42-46].

CONCLUSION

Supercapacitors are becoming ideal energy storage devices because of the inexpensive materials used during fabrication, long life and cycle time, ease of maintenance, availability, and energy densities. Natural sources are promising options to produce activated carbon for supercapacitor applications. This study shows that the activated carbons produced from the precursors meet the requirement as electrode material for such applications.

Researchers have found that biomass waste such as coconut shell, pistachio shell, acorn shell, bamboo, sugarcane bagasse, rice bran, algae, and areca fiber are excellent sources of producing activated carbons after stabilization, carbonization, and activation processes. The various modification procedures are also low cost and facile, and have high commercial potential compared to other sources. This AC has high specific surface area, high specific capacitance, and high energy density. It is also useful in designing and producing a variety of novel activated carbons with promising applications in high-performance energy storage supercapacitor devices. This book chapter covers all the detailed information used for supercapacitors. This study greatly benefits students, engineers, scientists, and other participants working in supercapacitor technologies.

ACKNOWLEDGMENT

The authors gratefully acknowledge Wichita State University for financial and technical support for this study.

REFERENCES

[1] Kötz, R., & Carlen, M. (2000). Principles and applications of electrochemical capacitors. *Electrochimica Acta*, *45*(15), 2483-2498.

[2] Lai, Y. Q., Li, J., Song, H. S., Zhang, Z. A., Li, J., & Liu, Y. X. (2007). Preparation of activated carbons from mesophase pitch and their electrochemical properties. *Journal of Central South University of Technology*, *14*, 633-637.

[3] Sunanda Tiwari, D. P., Sharma, D. N., & Kumar, R. T. S. (2013). Sapindus-based activated carbon by chemical activation. *Research Journal Material Science*, *1*, 9-15.

[4] Musolino, V., & Tironi, E. (2010, October). A comparison of supercapacitor and high-power lithium batteries. In *IEEE 2010*

Electrical Systems for Aircraft, Railway and Ship Propulsion (ESARS), pp. 1-6.

[5]   Signorelli, R., Ku, D. C., Kassakian, J. G., & Schindall, J. E. (2009). Electrochemical double-layer capacitors using carbon nanotube electrode structures. *Proceedings of the IEEE, 97*(11), 1837-1847.

[6]   Conway, B. E. (1999). *Electrochemical Supercapacitor: In Scientific Fundamentals and Technological Applications.* Kluwer Academic/ Plenum Publishers, New York.

[7]   Kularatna, N. (2014). *Energy Storage Devices for Electronic Systems: Rechargeable Batteries and Supercapacitors*, pp. 6-11, 149-180. Academic Press.

[8]   Rightmire, R. A. (1966). *U.S. Patent No. 3,288,641.* U.S. Patent and Trademark Office, Washington, DC.

[9]   Chmiola, J., Yushin, G., Gogotsi, Y., Portet, C., Simon, P., & Taberna, P. L. (2006). Anomalous increase in carbon capacitance at pore sizes less than 1 nanometer. *Science, 313*(5794), 1760-1763.

[10] Wu, F. C., Tseng, R. L., Hu, C. C., & Wang, C. C. (2005). Effects of pore structure and electrolyte on the capacitive characteristics of steam- and KOH-activated carbons for supercapacitors. *Journal of Power Sources, 144*(1), 302-309.

[11] Xia, Xinhui, et al. (2012). Integrated photoelectrochemical energy storage: Solar hydrogen generation and supercapacitor. *Scientific Reports 2*, 981.

[12] Markoulidis, F., et al. (2012). High-performance supercapacitor cells with activated carbon/MWNT nanocomposite electrodes. *IOP Conference Series: Materials Science and Engineering, 40*(1), 012021. IOP Publishing.

[13] *Electric double-layer capacitors.* Retrieved February 21, 2015, from http://research.omicsgroup.org/index.php/Supercapacitor#cite_note-9.

[14] Namisnyk, A. M. (2003). A *survey of electrochemical supercapacitor technology*, Doctoral Dissertation, University of Technology, Sydney. Retrieved November 10, 2015, from http://services.eng.uts.edu.au/ cempe/subjects_JGZ/eet/Capstone%20thesis_AN.pdf.

[15] Jabbarnia, A., Khan, W. S., Ghazinezami, A., and Asmatulu, R. (2016) "Investigating the Thermal, Mechanical and Electrochemical Properties of PVdF/PVP Nanofibrous Membranes for Supercapacitor Applications," *Journal of Applied Polymer Science*, DOI: 10.1002/app.43707.

[16] Ferreira, B., & Van der Merwe, W. (2014). *The Principles of Electronic and Electromechanic Power Conversion: A Systems Approach.* John Wiley & Sons.

[17] Mittal, A. K., Jagadesh Kumar, M., & Nalwa, H. S. (2011). Electrochemical double-layer capacitors featuring carbon nanotubes. In *Encyclopedia of Nanoscience and Nanotechnology*, Vol. 13, No. 271, pp. 263-271. American Scientific Publishers.

[18] Coskun, T. (2014). *Investigating solid-state supercapacitors constructed with PVA/CNT nanocomposite electrolytes*, Master's Thesis, Wichita State University. Retrieved March 7, 2015, from http://soar.wichita.edu/handle/10057/10951.

[19] Jabbarnia, A., Khan, W. S., Ghazinezami, A., and Asmatulu, R. (2016) "Tuning the Ionic and Dielectric Properties of Electrospun PVdF/PVP Nanofibers with Carbon Black Nanoparticles for Supercapacitor Applications," *International Journal of Engineering Research and Applications*, Vol. 6, pp. 65-73.

[20] Ohshima, H., & Furusawa, K. (Eds.). (1998). *Electrical Phenomena at Interfaces: Fundamentals: Measurements, and Applications*, Vol. 76, pp. 88-93. CRC Press.

[21] Zhu, Yanwu, et al. (2011). Carbon-based supercapacitors produced by activation of graphene. *Science 332*(6037), 1537-1541.

[22] Hunter, R. J. (2013). *Zeta Potential in Colloid Science: Principles and Applications*, Vol. 2, pp. 21-40. Academic Press.

[23] Wei, L., & Yushin, G. (2012). Nanostructured activated carbons from natural precursors for electrical double layer capacitors. *Nano Energy*, *1*(4), 552-565.

[24] Kalyani, P., Anitha, A., & Darchen, A. (2013). Activated carbon from grass—A green alternative catalyst support for water electrolysis. *International Journal of Hydrogen Energy*, *38*(25), 10364-10372.

[25] Hou, J., Cao, C., Ma, X., Idrees, F., Xu, B., Hao, X., & Lin, W. (2014). From rice bran to high energy density supercapacitors: A new route to control porous structure of 3D Carbon. *Scientific Reports, 4*, 7260.

[26] Xu, J., Gao, Q., Zhang, Y., Tan, Y., Tian, W., Zhu, L., & Jiang, L. (2014). Preparing two-dimensional microporous carbon from Pistachio nutshell with high areal capacitance as supercapacitor materials. *Scientific Reports, 4*, 5545.

[27] Natalia, M., Sudhakar, Y. N., & Selvakumar, M. (2013). Activated carbon derived from natural sources and electrochemical capacitance of double layer capacitor. *Indian Journal of Chemical Technology, 20*(6), 392-399.

[28] Jabbarnia, A., and Asmatulu, R. (2015) "Synthesis and Characterization of PVdF/PVP-Based Electrospun Membranes as Separators for Supercapacitor Applications," *Journal of Material Science and Technology Research*, Vol. 2, pp. 43-51.

[29] Yang, C. S., Jang, Y. S., & Jeong, H. K. (2014). Bamboo-based activated carbon for supercapacitor applications. *Current Applied Physics, 14*(12), 1616-1620.

[30] Zequine, Camila, et al. High Per formance and Flexible Supercapacitors based on Carbonized Bamboo Fibers for Wide Temperature Applications. *Scientific Reports, 6* (2016).

[31] Ismanto, A. E., Wang, S., Soetaredjo, F. E., & Ismadji, S. (2010). Preparation of capacitor's electrode from cassava peel waste. *Bioresource Technology, 101*(10), 3534-3540.

[32] Jiang, S., Shi, T., Long, H., Sun, Y., Zhou, W., & Tang, Z. (2014). High-performance binder-free supercapacitor electrode by direct growth of cobalt-manganese composite oxide nansostructures on nickel foam. *Nanoscale Research Letters, 9*(1), 1-8.

[33] Xiang, C., Li, M., Zhi, M., Manivannan, A., & Wu, N. (2013). A reduced graphene oxide/Co 3 O 4 composite for supercapacitor electrode. *Journal of Power Sources, 226*, 65-70.

[34] Kim, C., Lee, J. W., Kim, J. H., & Yang, K. S. (2006). Feasibility of bamboo-based activated carbons for an electrochemical supercapacitor electrode. *Korean Journal of Chemical Engineering, 23*(4), 592-594.

[35] Jain, A., & Tripathi, S. K. (2014). Fabrication and characterization of energy storing supercapacitor devices using coconut shell based activated charcoal electrode. *Materials Science and Engineering: B*, *183*, 54-60.

[36] Dandekar, M. S., Arabale, G., & Vijayamohanan, K. (2005). Preparation and characterization of composite electrodes of coconut-shell-based activated carbon and hydrous ruthenium oxide for supercapacitors. *Journal of Power Sources*, *141*(1), 198-203.

[37] Faisal, M. S. S., Rahman, M. M., & Asmatulu, R. (2016, April). Investigating effectiveness of activated carbons of natural sources on various supercapacitors. In *SPIE Smart Structures and Materials+ Nondestructive Evaluation and Health Monitoring* (pp. 980604-980604). International Society for Optics and Photonics.

[38] Md. Shahnewaz Sabit Faisal, *Studying activated carbons of natural sources for supercapacitor applications*, MS Thesis, Wichita State University, November 23, 2015.

[39] Hou, J., Cao, C., Ma, X., Idrees, F., Xu, B., Hao, X., & Lin, W. (2014). From rice bran to high energy density supercapacitors: A new route to control porous structure of 3D carbon. *Scientific Reports*, doi: 10.1038/srep07260.

[40] Si, W. J., Wu, X. Z., Xing, W., Zhou, J., & Zhuo, S. P. (2010). Bagasse-based nanoporous carbon for supercapacitor application. *Journal of Inorganic Materials* 26(1).

[41] Caturla, F., Molina-Sabio, M., & Rodriguez-Reinoso, F. (1991). Preparation of activated carbon by chemical activation with ZnCl2. *Carbon*, *29*(7), 999-1007.

[42] Chen, Xueyang, Kai Wu, Biao Gao, Qiaoyu Xiao, Jinhan Kong, Qi Xiong, Xiang Peng, Xuming Zhang, & Jijiang Fu. (2016). Three-dimensional activated carbon recycled from rotten potatoes for high-performance supercapacitors. *Waste and Biomass Valorization*, *7*(3), 551-557.

[43] Coskun, T., & Asmatulu, R. (2014). Enhancing the storage capacity of supercapacitors using PVA/CNT nanocomposite electrolytes, *ASME*

International Mechanical Engineering Congress and Exposition, Montreal, Canada, November 14-20, 7 pages.

[44] Nuraje, N., Asmatulu, R., & Mul, G. (2015). *Green Photo-Active Nanomaterials: Sustainable Energy and Environmental Remediation*, RSC Publishing, Cambridge, England.

[45] Asmatulu, R. (2013). *Nanotechnology Safety*, Elsevier, Amsterdam, The Nederland.

[46] Faisal, M.S.S., Rahman, M.M., and Asmatulu, R. (2016) "Investigating Effectiveness of Activated Carbons of Natural Sources on Various Supercapacitors," *SPIE Smart Structures/Non-Destructive Evaluation Conference*, Las Vegas, NV, March 20-24, 8 pages.

BIOGRAPHICAL SKETCHES

Md Shahnewaz Sabit Faisal

Education: Master of Science in Mechanical Engineering

Research and Professional Experience: 7 years.

Publications from the Last 3 Years:

Proceedings (Peer Reviewed)

1. Faisal, M. S. S., Downing, C., and Asmatulu, R. "Sealing the Holes of Aircraft Composites via Epoxy Nanocomposites Incorporated with Layered Nanoscale Inclusions," *CAMX Conference*, Anaheim, CA, September 26-29, 2016, 10 pages.

2. Faisal, M. S. S., Rahman, M., & Asmatulu, R. "Investigating Effectiveness of Activated Carbons of Natural Sources on Various Supercapacitors," *SPIE Smart Structures/Non-destructive Evaluation Conference,* Las Vegas, NV, March 20-24, 2016, 8 pages.

3. Faisal, M. S. S., Njoku, U., & Asmatulu, R. "Nanocomposite Sealants for Edge and Hole Treatment of Aircraft Carbon Fiber Composites," *CAMX/SAMPE Conference*, Dallas, TX, October 26-29, 2015, 14 pages.

4. Faisal, M. S. S., Ye, Z., Chen, Z., & Asmatulu, R. "Electrical Properties of Nanoscale Metallic Thin Films on Dielectric Elastomer at Various Strain Rates," *SPIE Smart Structures/Non-destructive Evaluation Conference*, San Diego, CA, April, 2015, 7 pages.

5. Ye, Z., Faisal, M. S. S., Asmatulu, R., & Chen, Z. "Bio-inspired Artificial Muscle Structure for Integrated Sensing and Actuation", *SPIE Smart Structures/Non-destructive Evaluation Conference*, San Diego, CA, April, 2015, 10 pages.

6. Faisal, M. S. S., Irving, J. L., Jurak, E., Jurak, S., and Asmatulu, R. "Highly Conductive Carbon Fibers and Nanoparticles for Hydrogen Production during Electrolysis Process," *CAMX/SAMPE Conference*, Orlando, FL, October 13-16, 2014, 8 pages.

7. Zhihang, Y., Faisal, M. S. S., Chen, Z. and Asmatulu, R. "Artificial Muscles of Dielectric Elastomers Attached to Artificial Tendons of Functionalized Carbon Fibers," *10th Graduate Research and Scholarly Projects (GRASP) Symposium*, Wichita State University, Wichita, KS, April 25, 2014 (extended abstract published).

8. Ye, Z., Faisal, M. S. S., Asmatulu, R., and Chen, Z. "Artificial Muscles of Dielectric Elastomers Attached to Artificial Tendons of Functionalized Carbon Fibers," *SPIE Smart Structures/Non-destructive Evaluation Conference*, San Diego, CA, March 9-13, 2014, 9 pages.

9. Faisal, M. S. S., Shagor, R. M. R., Farid, S. I., and Asmatulu, R. "Training Engineering Students for Hydrogen Production using Nanoparticles and Carbon Fiber Reinforced Composite Electrodes," *ASEE Midwest Section Conference*, Fort Smith, AR, September 24-26, 2014, 8 pages.

10. Husain, M., Asmatulu, R., Asaduzzaman, A., Usta, A., and Faisal, M. S. S. "Invention of Nanotechnology Based Arsenic Treatment

and Pathogen Removal System," *Seminar for Worldwide Danger of Arsenic Poisoning in Water and Sustainable Solutions*, Wichita, KS, December 7, 2013.

Ramazan Asmatulu

Education: Professor at WSU

Research and Professional Experience: 27 years.

Professional Appointments: Since 2006

Honors: Boeing Global Engineering Professor

Publications from the Last 3 Years:

Journal Articles (Peer Reviewed)

1. Saeednia, L., Yao, L., Cluff, K., and Asmatulu, R. "Sustained Releasing of Methotrexate from Injectable and Thermosensitive Chitosan-Carbon Nanotube Hybrid Hydrogels Effectively Controls Tumor Cell Growth," *ACS Omega*, 2019 (in press).
2. Le, L. N., Uddin, M. N., Zhang, B., Nair, R., and Asmatulu, R. "Effects of Graphene Oxide Thin Films and Nanocomposite Coatings on Flame Retardancy and Thermal Stability of Aircraft Composites" *Journal of Engineering Materials and Technology*, 2019 (in press).
3. Asmatulu, R., Veisi, Z., Uddin, M. N., and Mahapatro, A. "Highly Sensitive and Reliable Electrospun Polyaniline Nanofiber Based Biosensor as a Robust Platform for COX-2 Enzyme Detections," *Fibers and Polymers*, 2019 (in press).
4. Swarna, V. S., Alarifi, I. M., Khan, W. A., and Asmatulu, R. "Enhancing Fire and Mechanical Strengths of Epoxy

Nanocomposites for Metal/Metal Bonding of Aircraft Aluminum Alloys," *Polymer Composites*, 2019 (in press).

5. Van, T. T., Abedin, F., Usta, A., and Asmatulu, R. "Polyurethane Nanocomposite Coating with Silanized Graphene and Hexagonal Boron Nitride as Nanoadditives for Improved Resistance against UV Degradation," *Journal of Composite Materials*, 2018, https://doi.org/10.1177/0021998318799402.

6. Alarifi, I. M., Khan, W. S., and Asmatulu, R. "Synthesis of Electrospun Polyacrylonitrile- Derived Carbon Fibers and Comparison of Properties with Bulk Form," *Plos One*, 2018, Vol. 13(8), pp. e0201345, https://doi.org/10.1371/journal.pone. 0201345.

7. Alarifi, I. M., Alharbi, A. R., Khan, W. S., Usta, A., and Asmatulu, R. "Water Treatment Using Electrospun PVC/PVP Nanofibers as Filter Medium," *International Journal of Material Science and Research*, Vol. 1(2), pp. 43-49, 2018.

8. Kumar, S. S. A., Uddin, M. N., Rahman, M. M., and Asmatulu, R. "Introducing Graphene Thin Films into Carbon Fiber Composite Structures for Lightning Strike Protection," *Polymer Composites*, 2018, https://doi.org/10.1002/pc.24850.

9. Kurt, H. I., Oduncuoglu, Yilmaz, N. F., M., Ergul, E., Asmatulu, R. "A Comparative Study on the Effect of Welding Parameters of Austenitic Stainless Steels Using Artificial Neural Network and Taguchi Approaches with ANOVA Analysis," *Metals*, Vol. 8, 13 pages, 2018.

10. Rahman, A. K. M. S., Mathur, V. and Asmatulu, R. "Effect of Nanoclay and Graphene Inclusions on the Low-Velocity Impact Resistance of Kevlar-Epoxy Laminated Composites," *Composite Structures*, Vol. 187, pp. 481-488, 2018.

11. Moniruddin, M., Ilyassov, B., Zhao, X., Smith, E., Serikov, T., Ibrayev, N., Asmatulu, R., and Nuraje, N. "Recent Progress on Perovskite Nanomaterials in Photovoltaic and Water Splitting Applications," *Materials Today Energy*, Vol. 7, pp. 246-259, 2018.

12. Mohammad, S., Uddin, M. N., Hwang, G., and Asmatulu, R. "Superhydrophobic PAN Nanofibers for Gas Diffusion Layers of Proton Exchange Membrane Fuel Cells for Cathodic Water Management," *International Journal of Hydrogen Energy*, 2017, https://doi.org/10.1016/j.ijhydene.2017.07.229.

13. Asmatulu, R., Khan, A., Adigoppula, V. K., and Hwang, G., "Enhanced Transport Properties of Graphene-Based, Thin Nafion® Membrane for Polymer Electrolyte Membrane Fuel Cells," *International Journal of Energy Research*, Vol. 42, pp. 508-519, 2017.

14. Ye, Z., Chen, Z., Asmatulu, R., and Chan, O. "Robust Control of Dielectric Elastomer Diaphragm Actuator for Human Pulse Signal Tracking," *Smart Materials and Structures*, Vol. 26, 12 pages, 2017.

15. Seraz, M. S., Uddin, M. N., Yang, S. Y., and Asmatulu, R. "Investigating Electrochemical Behavior of Antibacterial Polyelectrolyte-Coated Magnesium Alloys for Biomedical Applications," *Journal of Environmental Science and Engineering Technology*, Vol. 5, pp. 23-33, 2017.

16. Zhang, B., Soltani, S. A., Le, L. N., and Asmatulu, R. "Fabrication nd Assessment of a Thin Flexible Surface Coating Made of Pristine Graphene for Lightning Strike Protection," *Materials Science and Engineering: B*, Vol. 216, pp. 31-40, 2017.

17. Ceylan, M., Yang, S. Y., and Asmatulu, R. "Effects of Gentamicin-Loaded PCL Nanofibers on Growth of Gram Positive and Gram Negative Bacteria," *International Journal of Applied Microbiology and Biotechnology Research*, Vol. 5, pp. 40-51, 2017.

18. Saeednia, L., Yao, L., Berrendt, M., Cluff, K., and Asmatulu, R. "Structural and Biological Properties of Thermosensitive Chitosan-Graphene Hybrid Hydrogels for Sustained Drug Delivery Applications," *Journal of Biomedical Materials Research: Part A*, Vol. 105, pp. 2381-2390, 2017.

19. Ghazinezami, A., Khan, W. S., Jabbarnia, A., and Asmatulu, R. "Impacts of Nanoscale Inclusions on Fire Retardancy, Thermal Stability, and Mechanical Properties of Polymeric PVC

Nanocomposites," *Journal of Thermal Engineering*, Vol. 3, pp. 1308-1318, 2017.

20. Kececi, E., and Asmatulu, R. "Effects of Moisture Ingressions on Mechanical Properties of Sandwich Structured Fiber Composites for Aerospace Applications," *International Journal of Advanced Manufacturing Technology*, Vol. 88, pp. 459-470, 2017.

21. Khan, W. S., Ceylan, M., Jabbarnia, A., Saeednia, L., and Asmatulu, R. "Structural Investigations of Electrospun PAN Nanofibers Incorporated with Various Nanoscale Inclusions," *Journal of Thermal Engineering*, Vol. 3, pp. 1375-1390, 2017.

22. Alharbi, A., Alarifi, I. M., Khan, W. S., Swindle, A., and Asmatulu, R. "Synthesis and Characterization of Electrospun Polyacrylonitrile/Graphene Nanofibers Embedded with $SrTiO_3/NiO$ Nanoparticles for Water Splitting," *Journal of Nanoscience and Nanotechnology*, Vol. 17, pp. 1-9, 2017.

23. Usta, A., and Asmatulu, R. "Synthesis and Characterization of Electrically Sensitive Hydrogels Incorporated with Cancer Drugs," *Journal of Pharmaceutics & Drug Delivery Research*, 2016, http://dx.doi.org/10.4172/2325-9604.1000146.

24. Jabbarnia, A., Khan, W. S., Ghazinezami, A., and Asmatulu, R. "Investigating the Thermal, Mechanical and Electrochemical Properties of PVdF/PVP Nanofibrous Membranes for Supercapacitor Applications," *Journal of Applied Polymer Science*, 2016, DOI: 10.1002/app.43707.

25. Moulod, M., Jalali, A., and Asmatulu, R. "Biogas Derived from Municipal Solid Waste to Generate Electric Power through Solid Oxide Fuel Cells," *International Journal of Energy Research*, Vol. 40, pp. 2091-2104, 2016.

26. Jiang, J., Ceylan, M., Jai, T., Yao, L., Asmatulu, R., and Yang, S. Y. "Poly-ε-Caprolactone Electrospun Nanofiber Mesh as a Gene Delivery Tool," *AIMS Bioengineering*, Vol. 3, pp. 528-537, 2016.

27. Kurt, H. I., Oduncuoglu, M., and Asmatulu, R. "Wear Behavior of Aluminum Matrix Hybrid Composites Fabricated through Friction

Stir Welding Process," *Journal of Iron and Steel Research International*, Vol. 23, pp. 1119-1126, 2016.

28. Mallonee, E., Barkley, J., and Asmatulu, R. "Training Renewable Energy Systems to Midwestern College Students for Engineering Education and Improved Retention Rates," *Transactions on Techniques in STEM Education*, Vol. 2, pp. 4-10, 2016.

29. Alarifi, I. M., Khan, W. S., Rahman, A. K. M., Kostogorova-Beller, Y., and Asmatulu, R. "Synthesis, Analysis and Simulation of Carbonized Electrospun Nanofibers Infused Carbon Prepreg Composites for Improved Mechanical and Thermal Properties," *Fibers and Polymers*, Vol. 17, pp. 1449-1455, 2016.

30. Kurt, H. I., Dere, M., Asmatulu, R., Guzelbey, I., and Salman, S. "Investigating the Relationships between Structures and Properties of Al Alloys Incorporated with Ti and Mg Inclusions," *Journal of Engineering Materials and Technology*, Vol. 138, 6 pages, 2016.

31. Asmatulu, R., Shinde, M. A., Alharbi, A., and Alarifi, I. M. "Integrating Graphene and C_{60} into TiO_2 Nanofibers via Electrospinning Process for the Enhanced Energy Conversion Efficiencies," *Macromolecular Symposia*, Vol. 365, pp. 128-139, 2016.

32. Mahat, K. B., Alarifi, I. M., Alharbi, A., and Asmatulu, R. "Effects of UV Light on Mechanical Properties of Carbon Fiber Reinforced PPS Thermoplastic Composites," *Macromolecular Symposia*, Vol. 365, pp. 157-168, 2016.

33. Alarifi, I. M., Alharbi, A., Khan, W. S., Rahman, A. K. M. S., and Asmatulu, R. "Mechanical and Thermal Properties of Carbonized PAN Nanofibers Cohesively Attached to Surface of Carbon Fiber Reinforced Composites," *Macromolecular Symposia*, Vol. 365, pp. 140-150, 2016.

34. Alharbi, A., Alarifi, I. M., Khan, W. S., and Asmatulu, R. "Synthesis and Analysis of Electrospun $SrTiO_3$ Nanofibers with NiO_x Nanoparticles Shells as Photocatalysts for Water Splitting," *Macromolecular Symposia*, Vol. 365, pp. 246-257, 2016.

35. Chinni, G., Belachew, I., and Asmatulu, R. "Hands-On Training the Engineering Students on Biodiesel Production Using Waste Vegetable Oils," *Transactions on Techniques in STEM Education*, Vol. 1, pp. 37-45, 2016.

36. Alharbi, A., Alarifi, I. M., Khan, W. S., and Asmatulu, R. "Highly Hydrophilic Electrospun Polyacrylonitrile/Polyvinypyrrolidone Nanofibers Incorporated with Gentamicin as Filter Medium for Dam Water and Wastewater Treatment," *Journal of Membrane and Separation Technology*, Vol. 5, pp. 38-56, 2016.

37. Asmatulu, R., Diouf, D., Moniruddin, M., and Nuraje, N. "Enhanced Antiweathering Nanocomposite Coatings with Silanized Graphene Nanomaterials," *International Journal of Engineering Research and Applications*, Vol. 6, pp. 79-91, 2016.

38. Jabbarnia, A., Khan, W. S., Ghazinezami, A., and Asmatulu, R. "Tuning the Ionic and Dielectric Properties of Electrospun PVdF/PVP Nanofibers with Carbon Black Nanoparticles for Supercapacitor Applications," *International Journal of Engineering Research and Applications*, Vol. 6, pp. 65-73, 2016.

In: An Essential Guide ...
Editor: Luke Lewin

ISBN: 978-1-53615-047-6
© 2019 Nova Science Publishers, Inc.

Chapter 3

FIELDS OF APPLICATION OF ELECTRICAL RESISTIVITY METHOD ON EARTH SCIENCES AND ENVIRONMENT: AN OVERVIEW

*Edite Martinho**
CERENA, Instituto Superior Técnico,
Universidade de Lisboa, Lisbon, Portugal

ABSTRACT

The electrical resistivity method is a non-destructive and inexpensive geophysical technique that can be applied to a variety of geological targets at small and medium depths. The method allows a mapping of the internal structure of investigated medium. Initially applied to mining and groundwater exploration, new applications related to environmental and geotechnical projects, archeology and stone cultural heritage have emerged in the last three decades. The method is also widely used in downhole logging. This chapter presents a brief overview of application areas of this geophysical method and gives a comprehensive and practical pureview of the results obtained in each of these areas.

---

* Corresponding Author's E-mail: emar@mail.ist.utl.pt.

Keywords: electrical resistivity method, mining exploration, groundwater exploration, environment, engineering, cultural heritage, archaeology, hydrocarbon exploration

1. INTRODUCTION

The materials electrical resistivity can be measured in the field or in the laboratory. In the field, the electrical resistivity can be measured at earth's surface or in boreholes, to investigate subsurface conditions in a given area. Rock resistivity is related to various geological parameters such as the mineral and fluid content, porosity and degree of water saturation in the rock. As a non-invasive and inexpensive tool, the method has been widely used in mining exploration (e.g., Gallas, 2015; Moreira et al., 2016), exploration, quality and protection of groundwater (e.g., Kazakis et al., 2016; Hasan et al., 2018; Oliveira Braga et al., 2006), environmental studies (e.g., Bichet et al., 2016; Maurya et al., 2017), engineering (e.g., Buvat et al., 2014; Merritt et al., 2016; Crawford et al., 2018) and cultural heritage including archaeology studies (e.g., Lehmann and Krüger, 2011; Martinho et al., 2012; Simyrdanis et al., 2016; Fischanger et al., 2018).

The investigation of the spatial variation of the apparent resistivity can be carried out using several techniques based on the projection of these variables in profiles, maps, vertical electrical soundings (VES) and pseudosections. The first two techniques allow the detection of lateral variations in the electrical properties of the geological materials. The measurements are made with a fixed electrode spacing resulting in the knowledge of the geophysical parameters at a pseudo-depth. The vertical electrical soundings allow the measuring of "vertical" variations in the electrical properties in a given geological section. The data of the electrical soundings are obtained with several electrode spacings, generally, according to a geometric progression. The interpretation of the electric soundings in parallel profiles allows the approximation the three-dimensional distribution of the geoelectric parameters when the geological structure behaves sub-horizontally. Pseudosection (electrical resistivity imaging) combines sounding and profiling, so it can give information about the resistivity distribution as function of both depth and horizontal distance. Resistivity images are created by inverting hundreds to thousands of individual resistivity measurements to produce a graphical representation of the subsurface resistivity (Loke and Barker, 1996a; Loke and Barker, 1996b).

In some fields of application, due to ambiguities in interpretation, the electrical resistivity method is integrated with other geophysical methods. In mining exploration, the method is often used with induced polarization (IP) in the search for disseminated ore. In the study of landfills contamination plumes, the method is widely used with induced polarization (in clayey areas, the IP helps to differentiate contaminated areas - low resistivity and low IP effect, from clayey lenticulas - low resistivity and high effect IP) and with frequency domain electromagnetic methods (FDEM). In the engineering studies, it has been used with IP and ground penetrating radar (GPR).

The purpose of this chapter is to list the application fields of electrical resistivity method and give a comprehensive and practical overview of some of more recent results obtained in each of these fields. The data comes from the most relevant literature published, mainly, in the last two decades in scientific journal databases. A critical overview is also presented.

2. APPLICATION FIELDS OF RESISTIVITY SURVEYS

Electrical resistivity method is very often used in groundwater investigations since it enables rapid and accurate examinations of the underground aquifers in what concerns to their geometrical (depth and thickness, position of the water table) and hydraulic (porosity, hydraulic conductivity and transmissivity) characteristics, water quality and contamination risk, and groundwater direction and flow velocity (e.g., Kazakis et al., 2016; Goebel et al., 2017). Mapping of subsurface geological structures and delineation of weathered/fracture zones for potential exploitation of groundwater (e.g., Hasan et al., 2018) as well as estimation of groundwater levels and recharge sources (e.g., Yeh et al., 2015) are other areas where the method is widely applied.

Since 1980, environmental issues have led to the application of the method in the study of contamination plumes from landfills (e.g., Bichet et al., 2016) and hydrocarbon spills (e.g., Blondel et al., 2014). In this type of studies, the method is generally used in conjunction with other geophysical methods such as electromagnetic methods (EM) and induced polarization (IP). In the case of contamination plumes caused by hydrocarbon spills, the method is already used to discriminate microbial activity (growth and biosurfactant production) (e.g., Masy et al., 2016).

The electrical resistivity method, namely, electrical resistivity tomography (ERT) is also widely used in engineering field. Indeed, the method has proven effective in landslide investigations (e.g., Perrone et al., 2014), in determination of several soil properties (e.g., Chrétien et al., 2014), permafrost characterization (e.g., Léger et al., 2017), as a soil quality control test (compaction) (e.g., Mostafa et al., 2018) and detection of cavities either in karstic or in other geological environments (e.g., Gochenour et al., 2018).

According to Crawford et al., (2018), in regard to landslide investigations, the method provides information about "location of the failure zone, slope morphology (scarps, flanks, toe bulges), contrasts in lithology and soil types, moisture regimes, and changes in moisture over time." Among the soil properties whose information can be obtained through the method it takes into account the structure, water content, fluid composition and hydrological processes (Samouelian et al., 2005). The main advantage of applying electrical resistivity is that the information is obtained without digging, unlike traditional methods, since it is a non-invasive technique. The permafrost characterization can also be made using electrical resistivity. Thaw layer thickness, snow depth and ice-wedge characteristics, depth of permafrost base and structure of permafrost (e.g., Léger et al., 2017) are parameters that can be estimated with this method. Integrated with other geophysical methods, namely seismic, it is also possible to obtain information about the freeze-thaw processes and their implications on the physical and mechanical properties of permafrost (e.g., Wu et al., 2017).

In the field of subsurface geology, the detection of karstic voids constitutes the main subject of the application of the method. The resistivity images can provide a clear view of weathered soils, weak areas or karst voids and bedrock. The groundwater pathway in karstic environment can also be investigated (e.g., Gochenour et al., 2018).

As the method is non-invasive, as is desirable in cultural heritage studies, this is a field in which the electrical resistivity technique has great application, mainly in stone built cultural heritage. The electrical resistivity allows to evaluate the state of degradation of historical buildings and cultural artifacts and, therefore, it is perhaps the most suitable technique for the exploration of these structures and for determining spatial moisture distribution in stone or masonry monuments (e.g., Sass and Viles, 2006; Leucci et al., 2007; Martinho et al., 2012). Also, in archaeology, the electrical resistivity is one of the most commonly applied geophysical techniques. The technique is suitable in detecting walls, cavities and other structures at different depths (e.g., Al-Saadi et al., 2018; Leucci and Greco, 2012).

The electrical resistivity method has been traditionally used in metals exploration such as gold and sulphides. Nevertheless it has been integrated, in most cases, with the induced polarization due to the widespread dissemination of these metals in pores, veins or lodes, often filled by quartz and barite. The IP method indicates the strength of polarization effects experienced by ions in the vicinity of metallic grains present in rock. The presence of metallic minerals in rocks increases the IP effect. Nowadays, the method is already being used in the attempt to discriminate different types of quartz (Gallas, 2015).

In hydrocarbon exploration, the method is used in boreholes to measure resistivity as a function of depth, i.e., the so-called resistivity logs. In the oil industry, the resistivity logs are, generally, run simultaneously with other logs (sonic, self potential (SP), radioactivity) so that different parameters of the penetrated formations can be recorded. Resistivity logs are important for identification of hydrocarbons zones (qualitative interpretation) and for quantitative calculation of hydrocarbon saturation of a reservoir. They are also used to detect mud cake and to measure mud resistivity.

In engineering, electrical resistivity is also widely used to evaluate concrete performance, mainly for the assessment of cracks and their influence on the durability of concrete structures (e.g., Lataste et al., 2003; Pacheco et al., 2014; Medeiros-Junior and Lima, 2016). This issue will not be addressed in this chapter because it belongs to the construction area.

3. EXAMPLES OF THE MOST RECENT RESISTIVITY INVESTIGATIONS IN GEOSCIENCES AND ENVIRONMENT

3.1. Groundwater: Exploration, Quality and Protection

3.1.1. Exploration

Kazakis et al., (2016a) used resistivity data, namely, vertical electrical soundings (VES) to estimate the hydraulic characteristics (porosity, hydraulic conductivity and transmissivity) of the aquifer of Anthemountas basin located in northern Greece. The obtained results showed a strong correlation with the obtained data from pumping tests. From the results, the authors also established a relationship between the aquifer resistivity and the hydraulic conductivity, which allows the estimation of the aquifer's hydraulic characteristics from VES measurements in areas without pumping test data.

Asfahani (2016) also used the VES technique to calculate the hydraulic parameters (transmissivity and hydraulic conductivity) of a Quaternary aquifer in the semi-arid Khanasser valley area (Syria). In this case, the approach took into account only the groundwater salinity. VES measurements in locations where water samples existed were used "in order to calibrate and establish empirical relationships between (i) transverse resistance Dar-Zarrouck (TR) parameter and modified transverse resistance (MTR), and (ii) between MRT and transmissivity (T)." These relationships allowed the extrapolation of the transmissivity in VES locations where the resistivities values of the water samples were not available. The results obtained for the hydraulic conductivity are, according to the author, in "acceptable agreement" with the results of the pumping tests in the Khanasser valley.

2D ERT was used together with magnetic method in order to map the subsurface structures and delineate the weathered/fracture zones for groundwater potential exploration in a highly heterogeneous site in South China (Hasan et al., 2018). The results of the two geophysical methods were integrated with information obtained from boreholes located in the study area. This integrated approach allowed the definition of the geological model of the studied area and detection several weathered/fracture zones, where is higher the potential of exploration of groundwater.

3.1.2. Quality and Protection

Oliveira Braga et al., (2006) used resistivity method (VES) in refinery areas of PETROBRAS (Alberto Pasqualini Refinery - REFAP) near Canoas (State of Rio Grande do Sul, Brasil) in order to evaluate the quality and protection of groundwater. The study allowed building the hydrogeological model of the area. It was identified the position of the water table as well as two layers of sand corresponding to two aquifers. The sand layer below the water table corresponds to an unconfined aquifer while the other, located between two clay layers, corresponds to a confined aquifer.

The relationship between resistivity and the Dar Zarrouk parameters allowed to suggest zones for the installation of monitoring wells and to indicate zones in which the confined aquifer is protected and areas of probable risk of contamination.

In some studies of this nature, electrical resistivity can be used together with other geophysical methods. For example, in a uranium mine in Ranger (Território do Norte, Austrália), electrical (SP-Self-Potential, resistivity and IP-Induced Polarization) and electromagnetic (TEM-Time domain EM) methods were used to detect groundwater contamination and seepage from structures used to store ore processing tailings (Buselli and Lu, 2001). The authors concluded that the IP and SP methods are good in direct detection of seepage while resistivity and TEM methods provide information about geological structure relevant to the hydrogeological processes controlling any seepage.

ERT was used along Monterey Bay coast in central California, aiming to understand the spatial distribution of freshwater and saltwater in the subsurface, and factors controlling this distribution (Goebel et al., 2017). The resistivity imaging was used "to identify geological flow conduits and barriers such as palaeo-channels and faults, localized saltwater intrusion from individual pumping wells, infiltration zones of surface fresh and brackish water, and regions showing improvements in water quality due to management actions." The study revealed a complex pattern of saltwater intrusion along the Monterey Bay. The same electrical resistivity technique (ERT) was also used by Kazakis et al., (2016b) to determine the extent and the geometrical characteristics of seawater intrusion in the coastal aquifer of the eastern Thermaikos Gulf, Greece. The results of the electrical resistivity tomography were combined with hydrochemical data, which allowed the construction of the saline intrusion map in that area of study.

3.2. Environment

In the last two decades the electrical resistivity has been one of the most used geophysical methods in the study of contamination plumes from landfills (e.g., Vaudelet et al., 2011; Bichet et al., 2016; Maurya et al., 2017).In the study of these contamination plumes, more than one geophysical method is generally used. In some cases, frequency domain electromagnetic methods (FDEM) are used in conjunction with resistivity. This is a logical combination since the main advantage of FDEM is that large areas can be surveyed rapidly and inexpensively, whereas their main disadvantage is that they give limited information about conductivity variation with depth. On the other hand, the resistivity method, either VES or pseudosections, gives good information about variation of resistivity with depth but is expensive and time-consuming to use in profiling mode. Another geophysical method that is often used along with resistivity is induced polarization (IP). In the study of soil contamination from landfills, the method is important because it helps to discriminate contaminated areas from clayey areas. The IP method has also been used as an aid to delineate plumes caused by organic contaminants although the results show that sometimes the IP response increases (e.g., Schmutz et al., 2010) while other times it decreases (e.g., Li et al., 2001).

Vaudelet et al., (2011) used resistivity/IP methods, spectral induced polarization (SIP) soundings and DFEM to map contaminant plumes in an urban site located in the Paris Basin (France). The integrated approach with different geophysical methods was justified by the fact that these types of sites are contaminated with different pollutants. The joint interpretation of the results of the EM (conductivity map), geochemistry and lithology allowed the selection of sites for the realization of resistivity/IP profiles. Resistivity/IP profiles revealed three major anomalous zones and SIP soundings provided information on the nature of the contaminant. Based on the results of this research, the authors propose a research methodology for sites of this type (contaminated urban areas). Leachate plume extension caused by a landfill composed by two cells (one older and one more recent) was determined by Bichet et al., (2016) using 2D ERT. The technique allowed delimiting both at depth and laterally the leachate plume and showing leachate infiltrations into the substratum. These infiltrations are linked to a highly fractured substratum and traversed by numerous discontinuous groundwater flows both revealed by 2D ERT. The techniques 2D and 3D ERT were also used in the study of migration of leachate plume from a heavily contaminated landfill located at Denmark (Maurya et al., 2017). The ERT results (Figure 1) were crossed with chemical data and showed a good correlation with leachate ionic strength. The variation in low resistivities was related with differences in the ionic strength of the landfill leachate.

Blondel et al., (2014) used resistivity and IP (Induced Polarization) methods to evaluate the temporal evolution of a hydrocarbon contamination plume caused by a hydrocarbon spill that occurred in 2009 due to a petroleum pipe breakdown. Laboratory and field measurements were carried out. The laboratory measurements were performed in fresh and biodegradated oil samples, both taken from a contaminated site. The field measurements were realized in 2009 (resistivity) and 2012 (resistivity and IP). The resistivity measured in the field over the contaminated area, between 2009 and 2012, showed a clear decrease related with oil biodegradation.

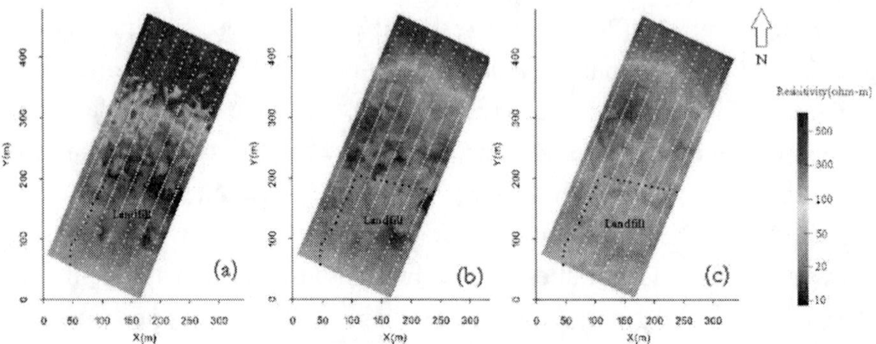

Figure 1. 3D DC inversion result shown as horizontal resistivity map at different elevation. (a) – 30 m, (b) – 20 m, (c) – 10 m. Boundary of the landfill is showen by ablack dotted line and the position of the electrodes is shown as white dots. (After Maurya et al., 2017).

Therefore, the resistivity parameter was a good indicator of the presence of partially degraded oil. On the other hand, the chargeability was important in the characterization of the geometry of the clayey-silt roof. ERT was also used by Masy et al., (2016) to monitor the biodegradation of hydrocarbons with *Rhodococcus erythropolis* T902.1 at a pilot scale. The ERT images allowed discriminating, successively, the lithological heterogeneities and microbial activities (growth and biosurfactant production). The authors concluded that, in the future, the technique should be more and more transferred to the field in order to monitor biodegradation processes and assist in selecting the most appropriate remediation technique.

3.3. Engineering

The electrical resistivity, mainly ERT, can be used to obtain information/detailed images of several soil properties (i.e., structure, water content, or fluid composition and hydrological processes) without digging (Samouelian et al., 2005) unlike traditional methods. In 2005, Ayuk et al.,

(2005) used the electrical resistivity method to study the foundations of part of the Federal Polytechnic Ado-Ekiti campus (South-Western Nigeria). The survey included VES and horizontal profiling from which it was possible to provide information about the nature, type and competence of the underlying material. The results allowed to construct geoelectric sections with the main geological units and to determine the respective properties (resistivities and thicknesses), as well as appropriate deep foundation design for the place (multi storey building). Buvat et al., (2014) studied the spatial variability of the soil through data of apparent resistivity. The approach used allowed them to discriminate between layers (detect the presence of a layer of clay) and the depth of the soil profile.

Understanding the factors influencing soil movements, especially moisture variations, it is of crucial importance in predicting problems that may occur at the foundations. Chrétien et al., (2014) used simultaneously ERT and time domain reflectometry (TDR) for qualifying the moisture distribution of a heterogeneous clay formation. The authors compared the resistivity variations obtained with local effective rainfall and soil moisture measurements from TDR down to a depth of 3 m. They concluded that the developed approach can be applied to heterogeneous field sites where a quantitative assessment of 2D soil moisture distribution is not possible. One of the main advantages of ERT is to allow the study to be done non-invasively. The importance of this issue led Merritt et al., (2016) to perform resistivity measurements in the laboratory aiming to develop resistivity-moisture content relationships of soils associated with a clayey landslide. They used different soils and different moisture contents, and also studied the influence of anisotropy. The experimental data were adjusted to petrophysical models. Model fit varied widely but the fit was better for clay dominated samples and worse for sand-dominated samples.

The electrical resistivity method is also being used as a soil quality control test. Mostafa et al., (2018) carried ou laboratory tests on calcareous soil samples for determining their electrical resistivity after compaction. Relationships between electrical resistivity and different soil properties, water content and compaction index were developed. The results showed that change in water and fines content are reflected on the electrical resistivity and therefore, in-situ compaction control can be made through electrical resistivity.

Landslide investigations are, in fact, a subject in which the electrical resistivity method is also been widely used since it can provide information regarding the type of landslide, location of the failure zone, slope morphology, differentiating soil and bedrock interfaces and moisture regimes (e.g., Perrone et al., 2014). For example, Crawford et al., (2018) used 2D ERT to predict the shear strength at two shallow colluvial landslides in Kentucky (USA). Electrical resistivity surveys were performed at different dates in the same location, which allowed determining the resistivity variations at time intervals, within the slope, associated to changes in hydrogeological conditions. Shear strength was calculated from ERT using a field-based framework that correlated shear strength and hydrologic parameters. The 2D resistivity models were used to interpret landslide failure zones, characteristics of landslide type, lithologic boundaries, soil thickness, and changes in moisture conditions (Figure 2).

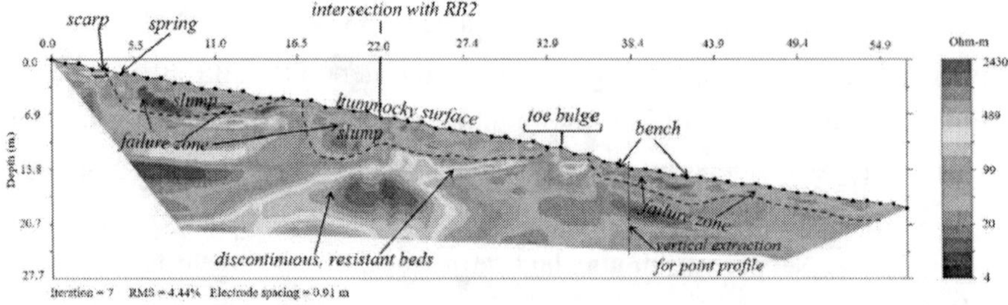

Figure 2. Electrical resitivity tomography profile of RB1. (After Crawford et al., 2018).

Observed changes in shear strength at depth in both landslides were associated to failure zones, specific soil horizons and areas of low resistivity, which allowed obtaining a spatial view of shear strength throughout the slope.

The electrical resistivity can also be used to characterize permafrosts. Léger et al., (2017) used ERT and GPR (ground penetrating radar) to characterize Arctic soil and permafrost near the village of Barrow on the Alaskan Arctic Coastal Plain, USA. They used the GPR to estimate thaw layer thickness, snow depth and ice-wedge characteristics. The structural information extracted from the GPR data was compared with ERT information about ice-wedge characteristics. The authors concluded that "GPR data collected during the frozen season can provide reliable estimates of active layer thickness and geometry of ice wedges" and that "the ice-wedgegeometry extracted from GPR data collected during the frozen season is consistent with ERT data." The GPR data were used to constrain the ERT inversion.

Saline permafrost soil during freeze-thaw transition was studied in laboratory by Wu et al., (2017) using spectral induced polarization (SIP) and seismic data aiming to understand the physical and mechanical processes during freeze-thaw cycles. The results contributed to a better petrophysical understanding of the physical and mechanical properties of saline permafrost during the freeze-thaw transitions but suggest a great uncertainty in the estimation of the unfrozen water content using electrical resistivity data.

As far as the study of subsurface geology is concerned, the method has also great application in the detection of cavities, either in the karstic environment (e.g., Farooq et al., 2012; Gochenour et al., 2018) or in other geological environments such as granite, phyllite and sandstone (e.g., Martínez-López et al., 2013) or in urban areas (e.g., Ungureanu et al., 2017). Farooq et al., (2012) used 2D and 3D ERT for detection karstic voids in a projected road in Yongweol-ri, South Korea. The resistivity images provided a clear view of weathered soils, weak areas or karst voids and bedrock. Gochenour et al., (2018) also used 2D ERT to study the connection between superficial features and karstic subsurface in two sites on Cave Hill at the Grand Caverns Natural National Landmark near Grottoes, Virginia (USA). The groundwater pathway has also been investigated. The results concluded that bedding geometry and rock type are the dominant factors that define both groundwater distribution and karstic features within Cave Hill. Martínez-López et al., (2013) used ERT to analyze the resolution of technique under different geological conditions and assess their potential for locating cavities. For that purpose, they analyzed the geoelectric response produced by three cavities located in different geological substrates (granite, phyllite and sandstone) previously characterized by direct methods. They also studied a mining void excavated in granite. According to the authors, the obtained results suggest that ERT is a viable technique for the detection and monitoring of mining voids and other subsurface cavities. Ungureanu et al., (2017) investigated the presence of underground voids on a construction site, in an urban area, with ERT. They used different electrode arrays whose pseudosections show the presence of underground voids. However, the position and size could not be accurately determined.

3.4. Cultural Heritage

In 1993, Haelterman et al., (1993) initiated a research program aiming to evaluate the capability of the electrical resistivity method to analyse the internal structure of masonry and the changes induced by grout injection on consolidation works of historical masonry. As the technique cannot be used without adaptations, they started studying the influence of the geometrical boundaries, electrode configuration and moisture content in the measurements. They concluded that the method could become a good tool for the design and control of the injection technique. Similarly, Van Gemert et al., (1996) and Venderickx and Gemert (2000) developed computer programs where the effect of the geometrical boundaries was incorporated, so that the resistivity maps were a picture of the real anomalies in the structure. They used these data processing techniques to visualize the internal structure of a thin wall surrounding the park of the Castle of Arenberg in Heverlee (Belgium) and the walls of the building "Duke's Mills (13th Century) at Aarschot (Belgium). Schueremans et al., (2003) used resistivity measurements for monitoring crack development and sealing done by cement grout injection on an experimental wall. Resistivity measurements were done in the sound, cracked and repaired situation. Mapping of the relative difference of resistivity values showed a clear picture of the masonry internal situation, showing therefore the applicability of the method in this field. As the moisture content is one of the factors that influence the resistivity, Sass and Viles (2006) successfully used the 2-D electrical resistivity method to investigate the walls moisture content of two abbey ruins.

Aiming to investigate the degradation degree of brick walls along a canal located in Venice (Italy), an ERT test was performed above and below m.s.l. along a portion of a canal's wall (Abu-Zeid and Santarato, 2006). From the results it was possible to recognize the degradation state where the visual inspection was impossible, which was useful for the restoration project. Similar work was performed by Abu-Zeid et al., (2010) on selected wall portions of the historical church of Montepetriolo, Perugia, Central Italy. 3D resistivity distribution models were obtained before and after grouting in order to assess the possibility of achieving a quantitative estimate of the injected mortar volumes to increase the mechanical resistance of the studied wall portions. In the most damaged marble column of the Crypt of the Cathedral of Otranto (Italy), Leucci et al., (2007) used ERT to study the presence of fractures within the column. The columns of the crypt were also investigated with GPR. The authors concluded that although both techniques are in good agreement, 2D electrical resistivity tomography seems to be less resolute than GPR because the results show a longitudinal fracture larger than its real dimension. However, the ERT results show clearly the wet zones of the column, coincident with the visible phenomena of biodeterioration, decomposition and loosening of the tight fabric of the mineral. Viles et al., (2008), Sass and Viles (2010 a, b) also used 2-D resistivity to provide information regarding moisture content in historical masonry walls in Oxford, England. Their work has proved ERT to be a non-destructive, inexpensive and effective tool for assessing moisture in stone monuments. The results gave preliminary confirmation of a simple model of catastrophic decay and illustrated the complexity of moisture regimes in historic walls. Lehmann and Krüger (2011) used impedance measurements to monitor the capillary transport of brine in sandstone samples. The experiment was carried out in laboratory aiming to study the effects of water and brine flow on the electrical impedance. The salts mainly influenced the impedance during the drying phase in the rate of increase of the impedance. Full 3-D ERT technique was used by Martinho et al., (2012) to obtain the moisture distribution patterns in three stone bas-reliefs located in the cloister of the Santa Cruz Monastery in Coimbra (Portugal).

Edite Martinho

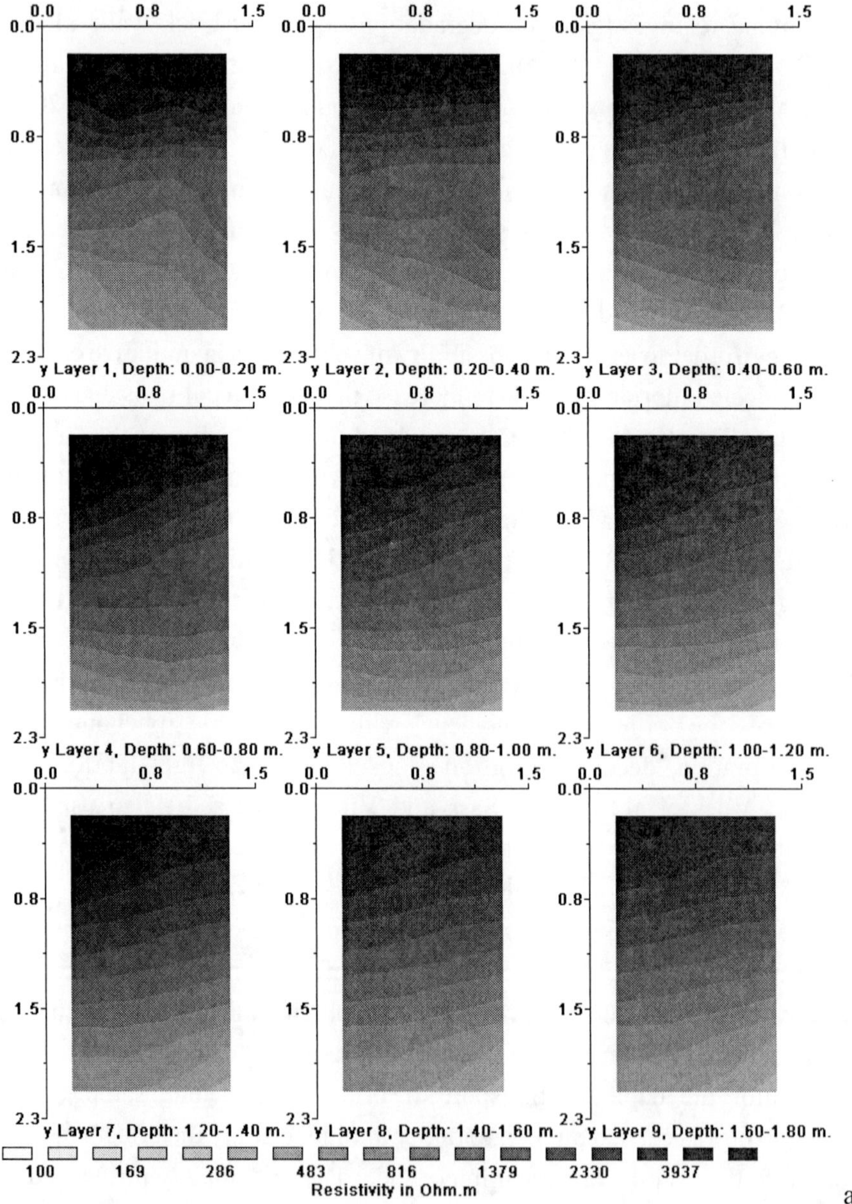

Figure 3. (Continued).

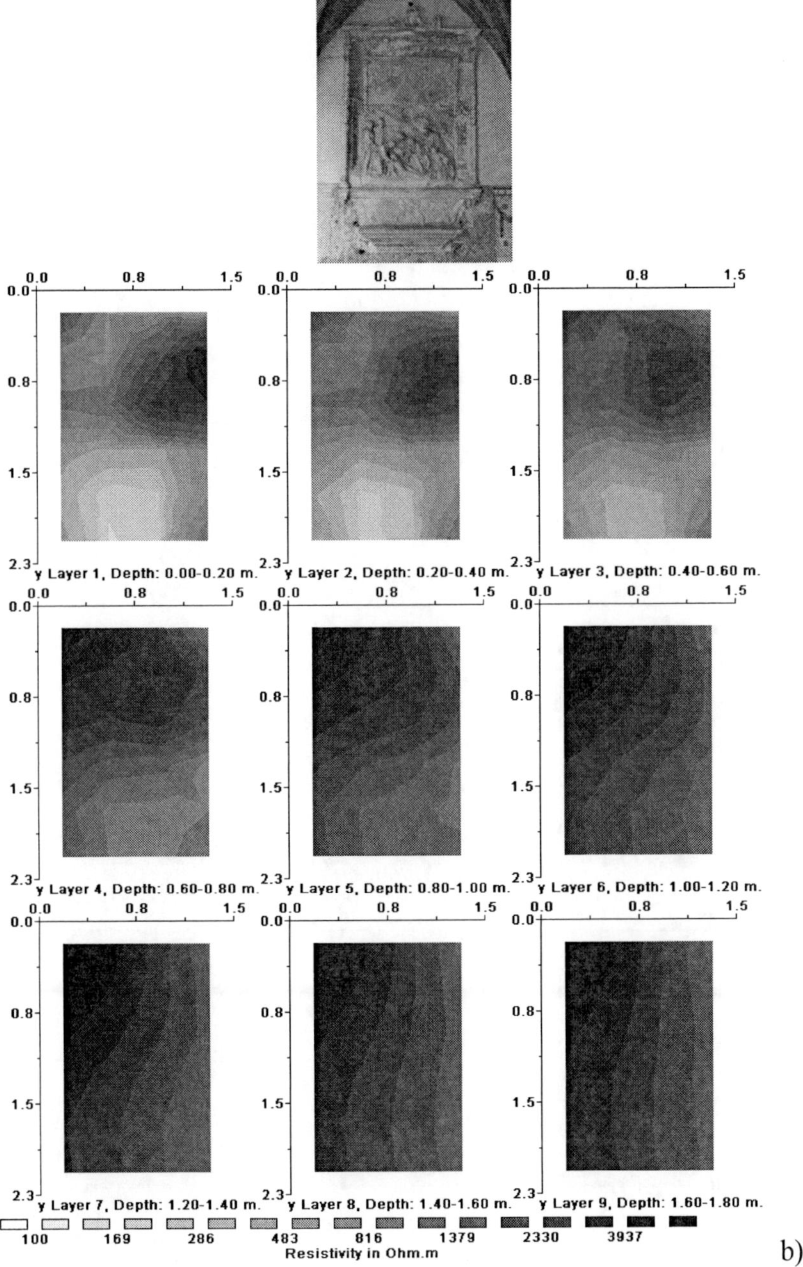

Figure 3. (Continued).

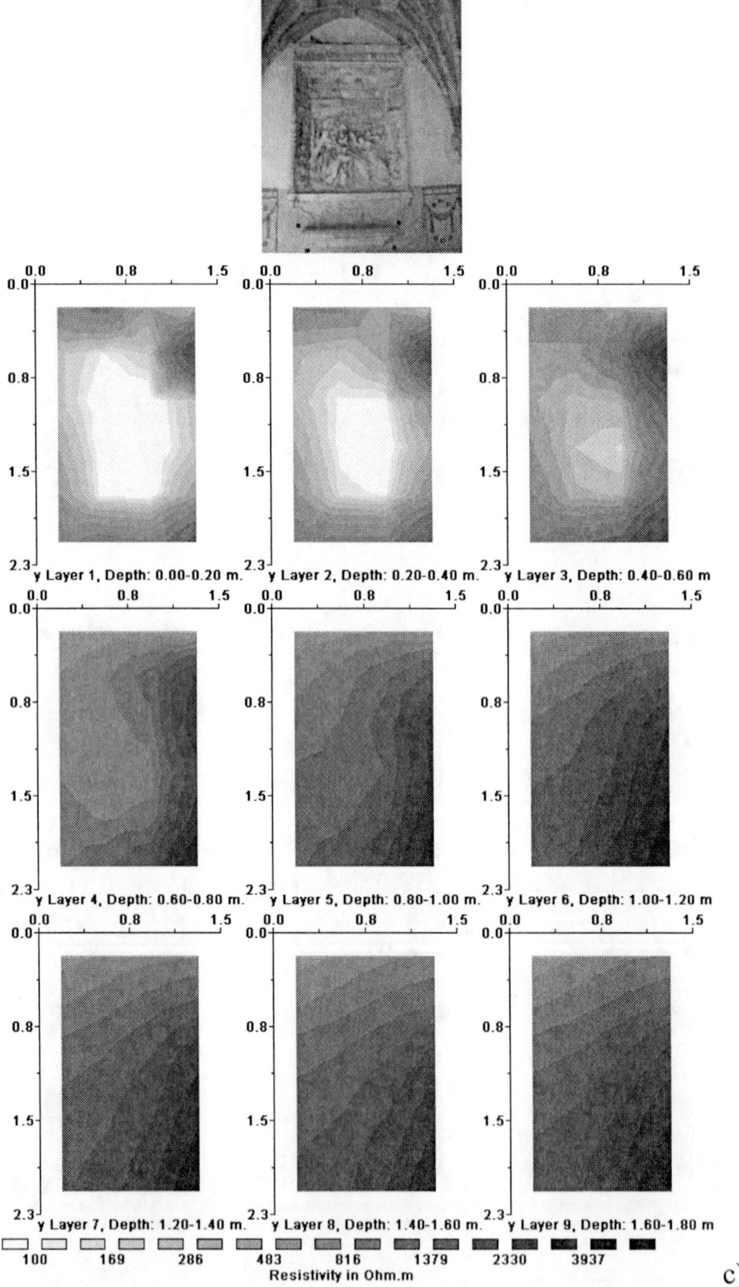

Figure 3. Resistivity sections in x-y plane generated from 3-D inversion of resistivity surveys. a) Masonry wall; b) PS panel; and c) PN panel. (After Martinho et al., 2012).

The bas-reliefs were sculpted in Pedra de Ançã (a soft, porous and light-colored limestone) by the sculptor Nicolau Chanterenne and have suffered severe deterioration from moisture and salts. The results of the full 3-D ERT technique were crossed with water-soluble salts data obtained from the panels. The moisture distribution patterns pointed to the existence of more than one source of moisture (Figure 3), which was consistent with data on water-soluble salts that suggested multiple saline sources contributing to the salt weathering on the bas-reliefs. The lower resistivity values observed in the panels also indicated severe decay.

Aiming to develope a methodology for monitoring the moisture on built heritage walls, Martínez-Garrido et al., (2018) used several integrated methods (non-or minimally destructive methods), among them ERT, on San Juan Bautista Church at Talamanca de Jarama (Spain). They concluded that, when the purpose is to distinguish between areas with widely varying moisture contents, ERT is more precise.

3.5. Archaeology

Ullrich et al., (2007) used ERT methods on Tell Jenderes and in a slag heap, in Northern Syria and Morocco, respectively. In Tell Jenderes, the resistivities values were correlated with different settlement phases from the Bronze Age from the Hellenistic period. In Morocco, the authors distinguished, using complex resistivity measurements, slag bodies from other archaeological features like walls, foundations and also from the surrounding bedrock. In both cases, archaeological objects were detectable at a depth of approximately 4 m below the surface. ERT imaging was also the technique chosen by Ortega et al., (2010) to prospect and analyse the morphology and geology subsurface of the Torcas-Cueva Mayor endokarst system (Sierra de Atapuerca, Spain). The technique allowed the identification of the endokarst morphologies and the main lithological groups. The study was important for understanding the distribution, in the area, of archaeological sites in order to plan their investigation. 3D ERT was undertaken by Leucci and Greco (2012) at the archaeological site of

Occhiolà (Sicily, Italy), inside the ruins of "Spirito Santo" Church. The survey aimed at exploring the buried structures, more precisely voids, related with possible crypt and/or tombs inside the church. The electrical surveys allowed locating anomalies that evidenced structures with regular form that can correspond to those structures (Figure 4).

In the last years, the ERT imaging has been used for locating onshore buried structures in archaeological sites as previously shown. However, the use of the technique in shallow offshore marine environments, important for studying the past dynamics in coastal areas, has been practically non-existent. Simyrdanis et al., (2016), used 2D and 3D ERT imaging to investigate part of the Hellenistic to Byzantine submerged archaeological site of Olous, located on the north-eastern coast of Crete, Greece. According to the authors, the ERT technique is suitable for locating submerged archaeological structures as walls, buildings and roads. Fernández-Álvarez et al., (2017) used the ERT technique combined with Ground Penetration Radar (GPR) as part of an archaeological project, in an area at the Pancorbo medieval site in Burgos (Spain), launched to study both the foundation of the castle and the possibility of a middle-aged village buried in its foot. 3D resistivity helped to identify anomalies present in the area, which were later verified through excavations showing a good correlation with the excavated remains. The buried foundations of a Roman villa near Nonnweiler (Germany) were also imaged by 3D ERT technique (Al-Saadi et al., 2018). The results allowed the characterization of the main structures of the Roman foundations (parts of the walls, pits and smaller internal structures of the Roman building). The anomalies caused by these structures presented a good correlation with the excavation results, either location or the vertical and horizontal extensions. Fischanger et al., (2018) also chose the ERT technique to investigate the area surrounding Tutankhamun's tomb with the objective of testing the hypothesis raised by Nicholas Reeves (Egyptologist), that Tutankhamun's tomb could be part of a larger tomb, possibly belonging to Queen Nefertiti.

The study revealed the presence of two anomalies located near Tutankhamun's funerary chamber. According to the authors, those anomalies do not seem to be related with known underground cavities. One may correspond to a void volume whose origin may be associated with Ancient Egypt, while the other, may be related with previous archaeological excavations. The authors did not reach a definitive conclusion on the origin of those anomalies.

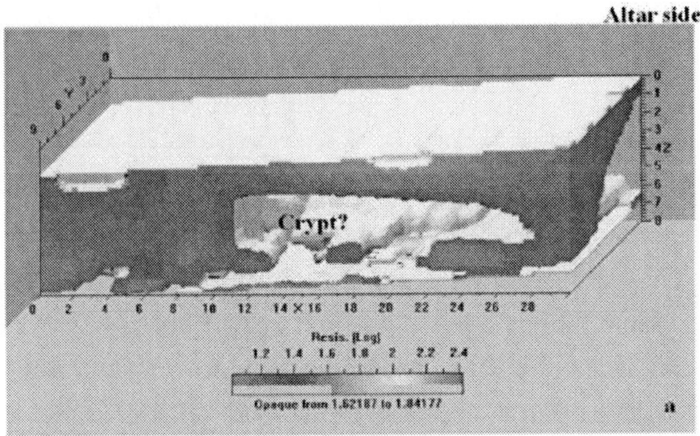

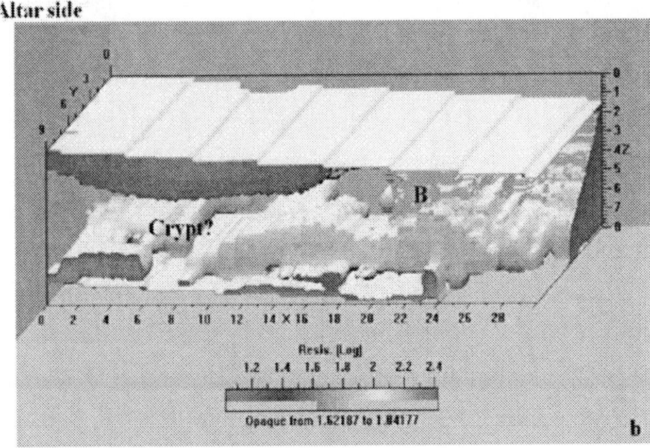

Figure 4. 3D contouring of iso-resistivity surfaces using a threshold value ranging from 40 to 65 Ω m. (After Leucci and Greco, 2012).

3.6. Mining Exploration

Gallas (2015) evaluated the ability of resistivity and IP methods in the detection of quartz masses of economic interest (hyaline high quality quartz or quartz with rutile inclusions) contained in rocks of granitic composition. The methods allowed to detect quartz masses (milky quartz masses, without economic interest) but did not allow distinguishing the mentioned types of quartz. The low resistivity anomalies, confirmed by excavations, were associated with quartz mass occurrences while high resistivity anomalies corresponded to non-fractured granites, without quartz masses. The author also concluded that in this type of exploration, the IP method does not allow to obtain results as satisfactory as the resistivity. The resistivity/IP techniques were also used by Moreira et al., (2016) to analyze the continuity of a mineralized quartz outcrop and the occurrence of new outcrops at one end of an inactive gold mine. The study was conducted in an area outside the mine. The results allowed the geophysical characterization of the mineralized zone (high resistivity and high chargeability) and the morphological mineralization modeling (Figure 5). A new lode in subsurface, located 30 m to the south of the lode outcrop was located. Then, given the results, the authors concluded that the resistivity/IP have the potential to be used as a criterion for sampling and defining of mineral content in exploration activities. Srigutomo et al., (2016) used 2D resistivity and IP measurements to delineate the presence of manganese ore in a scaly clay environment where manganese ores are hosted by mudstone. The zones identified as containing manganese ore were characterized by low resistivity (<5 ohm.m) and high chargeability (> 10 msec) values (Figure 6). However, according to the authors, the results need validation from borehole data.

With the purpose of imaging a possible area for mineral exploration, Bery et al., (2012) carried out a electrical resistivity and IP survey at an iron ore exploration area sit in Pagoh, Malaysia. The data allowed to define the geological structure and to find the iron ore (hematite). Interpretation was supported by geostatistical analysis performed from samples obtained locally. Yousefi et al., (2017) used electrical resistivity and IP methods in the prospecting of gold deposits in Meyduk (Latala), Kerman province, Iran, after the discovery of outcrops in the north of Meyduk. Two main veins were discovered but the authors only studied, one of them in more detail. The vein is located in direction N45E and was characterized by high chargeability and low resistivity, as expected.

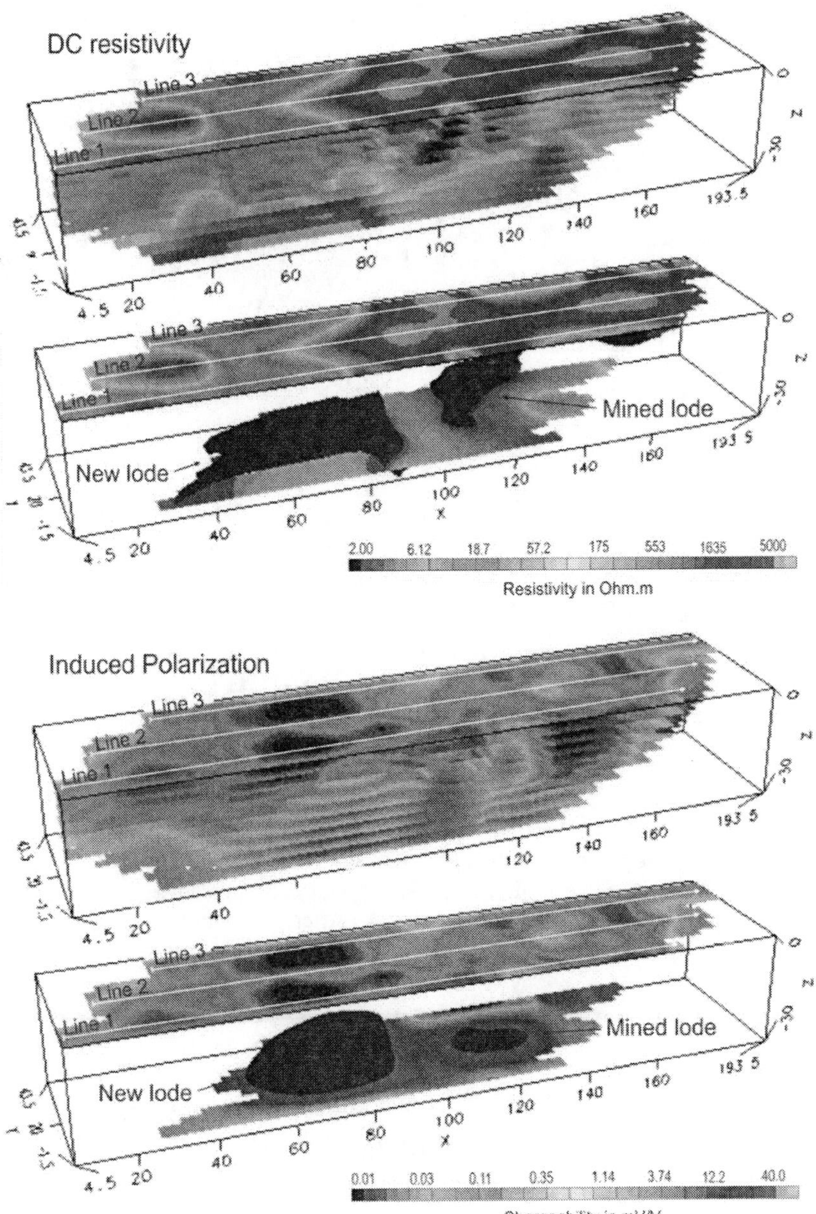

Figure 5. 3D visualization models for electrical resistivity and chargeability in slices and isosurface of 2500Ω.m and 30mV/V. (After Moreira et al., 2016).

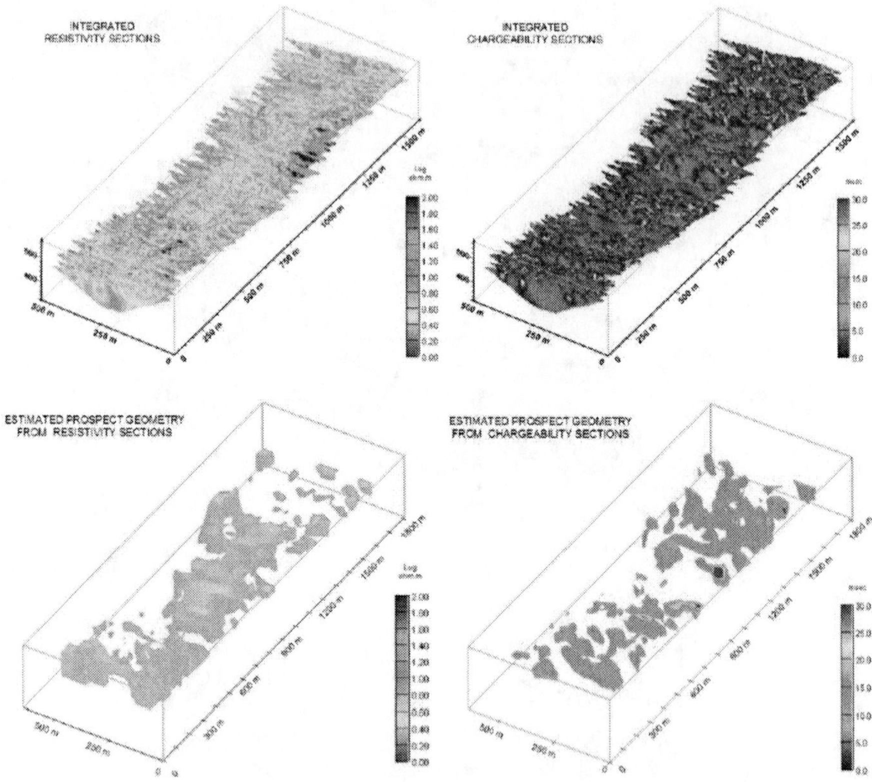

Figure 6. Arrangement of 2D resistivity sections in the surveyed area (upper left) and 3D interpolation of low resistivity bodies (lower left) based on the resistivity sections; chargeability sections (upper right) and 3D interpolation of high chargeability bodies based on the chargeability sections (lower right). (After Srigutomo et al., 2016).

3.7. Hydrocarbon Exploration

Integrated in one of the objectives of the Gulf of Mexico Gas Hydrate Joint Industry Project Leg II (GOM JIP Leg II), Collett et al., (2012) used a logging-while-drilling (LWD) data within gas-hydrate-bearing sand reservoirs in order to estimate the gas hydrates concentration under various geologic conditions. The data were obtained from electrical resistivity and acoustic logs aiming to examine the petrophysical nature of gas hydrate reservoirs and the distribution and concentration of gas hydrates within various complex reservoirs systems (Figure 7). The results confirmed that data obtained from these two types of logs can yield accurate gas hydrate saturation in systems such as sand reservoirs. However, it was not possible to have the same relationship in fractured reservoirs. In the development of a new technique for the accurate estimation of porosity in offshore gas hydrate occurrences in eastern India, Jain et al., (2018) also used LWD technology that included high-resolution resistivity imaging tool for fracture evaluation and depositional system identification.

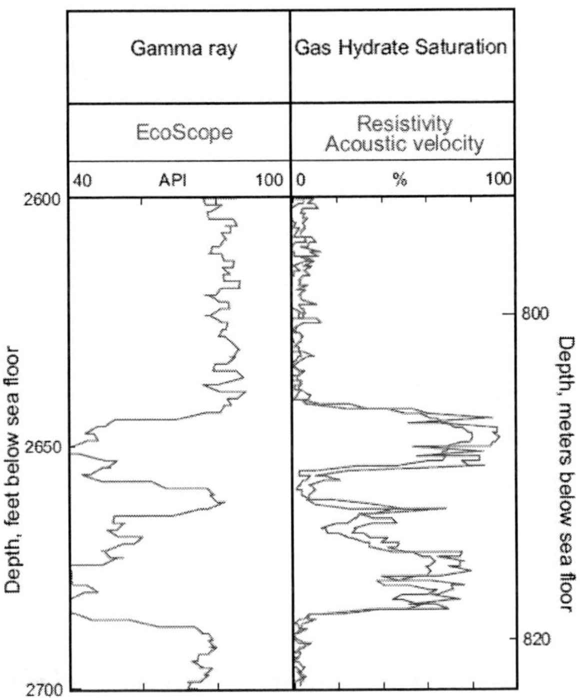

Figure 7. Gas hydrate saturations estimated from electrical resistivity and compressional-wave velocity logs in the Walker Ridge 313 H (WR313-H) well. (After Collett et al., (2012).

In oil and gas exploration, the resistivity method is routinely used in well logging given the great depths of the prospecting. However, there are reservoirs considered to be ultra-superficial (first 400 m depth) which are less expensive to exploit (Bauman, 2005). According to Bauman (2005), in these potential zones, 2-D electrical resistivity is an economical tool to investigate such reservoirs. Therefore, Bauman (2005) shows results of applying of the 2-D resistivity surveying in the exploration and direct detection of oil sands (in the Athabasca Oil Sand Deposits) and gas chargead quaternary channels, in Alberta, Canada.

4. CRITICAL OVERVIEW

Data from the literature published in the last two decades reveals that almost all published papers in peer review journals are based on field work. This is related to the fact that the electrical resistivity parameter, on which the method is based, has such great variability that it has long made it an attractive physical parameter for subsurface exploration and description of various geological targets at shallow to intermediate depths.

Regarding the techniques used to investigate the spatial variation of apparent resistivity, the electrical resistivity tomography (ERT) is the most employed followed by the vertical electrical soundings (VES). Electrical resistivity tomography is a recently developed geophysical technique used to obtain images of geological structures from electrical measurements made at the surface or in borehole. Resistivity images are created by inverting hundreds to thousands of individual resistivity measurements (Loke and Barker, 1996a, 1996b) to produce a graphical representation of the subsurface resistivity. The development of the technology associated with automatically multiplexed electrode arrangements and automatic measuring systems, that facilitate the acquisition of a large number of measurements in a limited time, with the use of automated resistivity inversion schemes, make the application of ERT a successful technique, even in complex geological environments. A 3-D electrical subsurface image can be obtained with 2-D or 3-D surveys. Since all geological structures are 3-D in nature, a fully 3-D resistivity survey using a 3-D interpretation model gives more accurate results. Therefore, the ERT technique is the most appropriate procedure and the most used technique to delineate the location and shape of the most probable sources of the anomalies detected at the ground surface.

As already previously mentioned, in several types of studies, this method is integrated with other geophysical methods, namely, with induced polarization method (IP), frequency domain electromagnetic methods (FDEM) and ground penetrating radar (GPR). In general, induced polarization is used to help solve ambiguities. In studies of contamination plumes from landfills, induced polarization is used to discriminate contaminated areas (low resistivity and low IP effect) from clayey areas (low resistivity and high IP effect), since most of these structures are implanted in unconsolidated sedimentary formations. In contamination plumes from hydrocarbon spills, induced polarization has been used as an aid in the delineation of the plumes and in the interpretation of the processes involved in the degradation of this type of compounds. This is a subject still under investigation since the results show that sometimes the IP response increases (e.g., Schmutz et al., 2010) while other times it decreases (e.g., Li et al., 2001). The frequency domain electromagnetic methods (FDEM) are also widely used with resistivity but at a prior stage to the use of resistivity, i.e., in the reconnaissance phase. These methods allow that large areas can be surveyed rapidly and inexpensible. However, they give limited information about variation of conductivity with depth unlike resistivity that gives good information about variation of resistivity with depth but is expensive and time-consuming to use in profiling mode. The GPR is another geophysical method that can also be used in the reconnaissance phase in studies of this nature.

In mining exploration, the induced polarization is mainly used with resistivity, in the search of disseminated ores. These type of ores gives rise to high IP effects due to electrode polarization.

In cultural heritage field, the GPR is sometimes used with resistivity as an aid in the characterization of fractures and location of humidity.

ACKNOWLEDGMENT

This work was supported by the Fundação para a Ciência e Tecnologia [Strategic project UID/ECI/04028/2019] to CERENA.

REFERENCES

Abu-Zeid, N., Santarato, G., 2006. Non-invasive imaging of ancient foundations status in Venice (Italy) using the Electrical Resistivity Tomography technique. *Heritage, Weathering and Conservation* – Fort, Alvarez de Buergo, Gomez-Heras &Vasquez-Calvo (eds), Taylor & Francis Group London ISBN 0-415-41272-2.

Abu Zeid, N., Balducci, M., Bartocci, F., Regni, R., Santarato, G., 2010. Indirect estimation of injected mortar volume in historical walls using the electrical resistivity tomography. *J. Cult Heritage* 11: 220-227.

Al-Saadi, O. S., Schmidt, V., Becken, M., Fritsch, T., 2018. Very-high-resolution electrical resistivity imaging of buried foundations of a Roman villa near Nonnweiler, Germany. *Archaeol Prospect* 25:209–218. DOI: 10.1002/arp.1703.

Asfahani, J., 2016. Hydraulic parameters estimation by using an approach based on vertical electrical soundings (VES) in the semi-arid Khanasser valleyregion, Syria. *J Afr Earth Sci* 117: 196-206.

Ayuk, A. M., Olayanju, G. M., Adelusi, A. O., 2005. Use of electrical resistivity technique for engineering site investigation: a case study from Ado-Ekiti, South-Western Nigeria. *Global Journal of Pure and Applied Sciences* 11: 534-552.

Bauman, P., 2005. 2-D resistivity surveying for hydrocarbons – a primer. *Recorder Journal* 30(4). https://csegrecorder.com/articles/view/2-d-resistivity-surveying-for-hydrocarbons-a-primer.

Bery, A. A., Saad, R., Mohamad, E. T., Jinmin, M., Azwin, I. N., Akip Tan, N. M., Nordiana, M. M., 2012. Electrical Resistivity and Induced Polarization Data Correlation with Conductivity for Iron Ore Exploration. *EJGE* 17: 3223-3337.

Bichet, V., Griseyb, E., Aleya, L., 2016. Spatial characterization of leachate plume using electrical resistivity tomography in a landfill composed of old and new cells (Belfort, France). *Eng Geol* 211: 61–73.

Blondel, A., Schmutz, M., Franceschi, M., Tichané, F., Carles, M., 2014. Temporal evolution of the geoelectrical response on a hydrocarbon contaminated site. *J Appl Geophys* 103: 161–171.

Buselli, G., Lu, K., 2001. Groundwater contamination monitoring with multichannel electrical and electromagnetic methods. *J Appl Geophys* 48: 11-23.

Buvat, S., Thiesson, J., Michelin, J., Nicoullaud, B., Bourennane, H., Coquet, Y., Tabbagh, A., 2014. Multi-depth electrical resistivity survey for mapping soil units within two 3 ha plots. *Geoderma* 232–234: 317–327.

Chrétien, M., Lataste, J. F., Fabre, R., Denis, A., 2014. Electrical resistivity tomography to understand clay behavior during seasonal water content variations. *Eng Geol* 169: 112–123.

Collett, T. S., Lee, M. W., Zyrianova, M. V., Mrozewski, S. A., Guerin, G., Cook, A. E., Goldberg, D. S., 2012. Gulf of Mexico Gas Hydrate Joint Industry Project Leg II logging-while-drilling data acquisition and analysis. *Mar Petrol Geol* 34: 41-61.

Crawford, M. M., Bryson, L. S., Woolery, E. W., Wang, Z., 2018. Using 2-D electrical resistivity imaging for joint geophysical and geotechnical characterization of shallow landslides. *J Appl Geophys* 157: 37–46.

Farooq, M., Park, S., Song, Y. S., Kim, J. H., Tariq, M., Abraham, A. A., 2012. Subsurface cavity detection in a karst environment using electrical resistivity (er): a case study from yongweol-ri, South Korea. *Earth Sci Res J* 16: 75-82.

Fernández-Álvarez, J. P., Rubio-Melendi, D., Castillo, J. A. Q., González-Quirós, A., Cimadevilla-Fuente, D., 2017. Combined GPR and ERT exploratory geophysical survey of the Medieval Village of Pancorbo Castle (Burgos, Spain). *J Appl Geophys* 144: 86-93.

Fischanger, F., Catanzariti, G., Comina, C., Sambuelli, L., Morelli, G., Barsuglia, F., Ellaithy, A., Porcelli, F., 2018. Geophysical anomalies detected by electrical resistivity tomography in the area surrounding Tutankhamun's tomb. *J Cult Heritage,* https://doi.org/10.1016/j.culher.2018.07.011.

Gallas, J. D. F., 2015. Quartz prospecting with induced polarization (IP) and resistivity by using gradient and dipole-dipole arrays. *Brazilian Journal of Geophysics* (RBGF) 33:555-564.

Gochenour, J., McGary, R. S., Gosselin, G., Suranovic, B., 2018. Investigating subsurface void spaces and groundwater in Cave Hill karst using resistivity. *15Th Sinkhole Conference, NCKRI Symposium*, pp 241-250.

Goebel, M., Pidlisecky, A., Knight, R., 2017. Resistivity imaging reveals complex pattern of saltwater intrusion along Monterey coast. *J Hydrol* 551: 746–755.

Haelterman, K., Lambrechts, A., Janssens, H., Van Gemert, D., 1993. Geo-electrical survey of masonry. *Mater Struct* 26: 495-499.

Hasan, M., Shang, Y., Jin, W., 2018. Delineation of weathered/fracture zones for aquifer potential using an integrated geophysical approach: A case study from South China. *J Appl Geophys* 157: 47–60.

Jain, V., Saumya, S., Vij, J., Singh, J., Singh, B., Pattnaik, S., Oli, A., Kumar, P., Collett, T.S., 2018. New technique for accurate porosity estimation from logging-while-drilling nuclear magnetic resonance data, NGHP-02 expedition, offshore, India. *Mar Petrol Geol,* https://doi.org/10.1016/j.marpetgeo.2018.11.001.

Kazakis, N., Vargemezis, G., Voudouris, K.S., 2016a. Estimation of hydraulic parameters in a complex porous aquifer system using geoelectrical methods. *Sci Total Environ* 550: 742–750.

Kazakis, N., Pavlou, A., Vargemezis, G., Voudouris, K. S., Soulios, G., Pliakas, F., Tsokas, G., 2016b. Seawater intrusion mapping using electrical resistivity tomography and hydrochemical data. An application in the coastal area of eastern Thermaikos Gulf, Greece. *Sci Total Environ* 543: 373–387.

Lataste, J. F., Sirieix, C., Breysse, D., Frappa, M., 2003. Electrical resistivity measurement applied to cracking assessment on reinforced concrete structures in civil engineering. *NDT&E Int* 36: 383–394.

Léger, E., Dafflon, B., Soom, F., Peterson, J., Ulrich, C., Hubbard, S., 2017. Quantification of Arctic Soil and Permafrost Properties Using Ground-Penetrating Radar and Electrical Resistivity Tomography Datasets. *IEEE Journal of Selected Topics in Applied Earth Observations and Remote Sensing* 10: 4348-4359.

Lehmann, F., Krüger, M., 2011. Wireless impedance measurements to monitor moisture and salt migration in natural stone. Cultural Heritage Preservation, *Proceedings of the European Workshop on Cultural Heritage Preservation* Germany Fraunhofer IRB Verlag ISBN: 978-3-8167-8560-6 pp 224-231.

Leucci, G., Cataldo, R., De Nunzio, G., 2007. Assessment of fractures in some columns inside the crypt of the Cattedrale di Otranto using integrated geophysical methods. *J Archaeol Sci* 34: 222-232.

Leucci, G., Greco, F., 2012. 3D ERT Survey to Reconstruct Archaeological Features in the Subsoil of the "Spirito Santo" Church Ruins at the Site of Occhiolà (Sicily, Italy). *Archaeology* 1: 1-6.

Li, C., Tercier, P., Knight, R., 2001. Effect of sorbed oil on the dielectric properties of sand and clays. *Water Resour Res* 37: 1783–1793.

Loke, M. H., Barker, R. D., 1996a. Pratical techniques for 3D resistivity surveys and data inversion. *Geophys Prospect* 44: 499-523.

Loke, M. H., Barker, R. D., 1996b. Rapid least-squares inversion of apparent resistivity pseudo-sections using quasi-Newton method. *Geophys Prospect* 44: 131-152.

Martínez-Garrido, M. I., Fort, R., Gómez-Heras, M., Valles-Iriso, J., Varas-Muriel, M. J., 2018. A comprehensive study for moisture control in cultural heritage using non-destructive techniques. *J Appl Geophys* 155: 36–52.

Martinez-López, J., Rey, J., Duenas, J., Hidalgo, C., Benavente, J., 2013. Electrical tomography applied to the detection of subsurface cavities. *J Cave Karst Stud* 75: 28–37.

Martinho, E., Alegria, F., Dionisio, A., Grangeia, C., Almeida, F., 2012. 3D-resistivity imaging and distribution of water soluble salts in Portuguese Renaissance stone bas-reliefs. *Eng Geol* 141-142: 33-44.

Masy, T., Caterina, D., Tromme, O., Lavigne, B., Thonart, P., Hiligsmann, S., Nguyen, F., 2016. Electrical resistivity tomography to monitor enhanced biodegradation of hydrocarbons with Rhodococcus erythropolis T902.1 at a pilot scale. *J Contam Hydrol* 184: 1–13.

Maurya, P. K., Rønde, V. K., Fiandaca, G., Balbarini, N., Auken, E., Bjerg, P. L., Christiansen, A. V., 2017. Detailed landfill leachate plume mapping using 2D and 3D electrical resistivity tomography - with correlation to ionic strength measured in screens. *J Appl Geophys* 138: 1–8.

Medeiros-Junior, R. A., Lima, M. G., 2016. Electrical resistivity of unsaturated concrete using different types of cement. *Constr Build Mater* 107: 11–16.

Merritt, A. J., Chambers, J. E., Wilkinson, P. B., West, L. J., Murphy, W., Gunn, D., Uhlemann, S., 2016. Measurement and modelling of moisture—electrical resistivity relationship of fine-grained unsaturated soils and electrical anisotropy. *J Appl Geophys* 124: 155–165.

Moreira, C. A., Borssatto, K., Ilha, L. M., Fernandes dos Santos, S., Rosa, F. T. G., 2016. Geophysical modeling in gold deposit through DC resistivity and induced polarization methods. REM – *Int Eng J* 69: 293-299.

Mostafa, M., Anwar, M. B., Radwan, A., 2018. Application of electrical resistivity measurement as quality control test for calcareous soil. *Housing and Building National Research Center HBRC Journal* 14: 379-384.

Oliveira Braga, A. C., Malagutti Filho, W., Dourado, J. C., 2006. Resistivity (DC) method applied to aquifer protection studies. *Brazilian Journal of Geophysics* 24: 573-581.

Ortega, A. I., Benito-Calvo, A., Porres, J., Pérez-Gonzalez, A., Martin Merino, A., 2010. Applying Electrical Resistivity Tomography to the identification of endokarstic geometries in the Pleistocene sites of the Sierra de Atapuerca (Burgos, Spain). *Archaeol Prospect* 17: 233–245.

Pacheco. J., Šavija, B., Schlangen, E., Polder, R. B., (2014). Assessment of cracks in reinforced concrete by means of electrical resistance and image analysis. *Constr Build Mater* 65: 417–426.

Perrone, A., Lapenna, V., Piscitelli, S., 2014. Electrical resistivity tomography technique for landslide investigation: A review. *Earth-Science Reviews* 135: 65-82.

Samouelian, A., Cousin, I., Tabbagh, A., Bruand, A., Richard, G., 2005. Electrical resistivity survey in soil science: a review. *Soil Till Res* 83: 173–193.

Sass, O., Viles, H. A., 2006. How wet are these walls? Testing a novel technique for measuring moisture in ruined walls. *J Cultur Heritage* 7: 257-263.

Sass, O., Viles, H. A., 2010a. Wetting and drying of masonry walls: 2-D resistivity monitoring of driving rain experiments on historic stonework in Oxford, UK. *J Appl Geophys* 70: 72-83, DOI: 10.1016/j. jappgeo.2009.11.006.

Sass, O., Viles, H. A., 2010b. Two-dimensional resistivity surveys of the moisture content of historic limestone walls in Oxford, UK: implications for understanding catastrophic stone deterioration. Limestone in the Built Environment: *Present-Day Challenges for the Preservation of the Past.* Geological Society London Special Publications 331 pp 237-249 DOI: 10.1144/SP331.22.

Schmutz, M., Revil, A., Vaudelet, P., Batzle, M., Femenia Vinao, P., Werkema, D., 2010. Influence of oil saturation upon spectral induced polarization of oil bearing sands. *Geophys J Int* 183: 211–224.

Schueremans, L., Rickstal, F., Venderickx, K., Van Gemert, D., 2003. Evaluation of masonry consolidation by geo-electrical relative difference resistivity mapping. *Mater Struct* 36: 46-50.

Simyrdanis, K., Papadopoulos, N., Cantoro, G., 2016. Shallow Off-Shore Archaeological Prospection with 3-D Electrical Resistivity Tomography: The Case of Olous (Modern Elounda), Greece. *Remote Sens* 8: 897, https://doi.org/10.3390/rs8110897.

Srigutomo, W., Trimadona, Pratomo, P. M., 2016. 2D Resistivity and Induced Polarization Measurement for Manganese Ore Exploration. *J Phys Conf Ser* 739 (2016) 012138: 1-6.

Ullrich, B., Günther, T., Rücker, C., 2007. Electrical Resistivity Tomography Methods for Archaeological Prospection. In: Posluschny, A., Lambers, K., Herzog, I. (Hrsgg.): Layers of Perception. *Proceedings of the 35th International Conference on Computer Applications and Quantitative Methods in Archaeology* (CAA), Berlin, 2.-6. April 2007, doi: 10.11588/propylaeumdok.00000488.

Ungureanu, C., Priceputu, A., Bugea, A. L., Chirică, A., 2017. Use of electric resistivity tomography (ERT) for detecting underground voids on highly anthropized urban construction sites. *Procedia Engineering* 209: 202–209.

Van Gemert, D., Janssens, H., Van Rickstal, F., 1996. Evaluation of electrical resistivity maps for ancient masonry. *Mater Struct* 29: 158-163.

Vaudelet, P., Schmutz, M., Pessel, M., Franceschi, M., Guérin, R., Atteia, O., Blondel, A., Ngomseu, C., Galaup, S., Rejiba, F., Bégassat, P., 2011. Mapping of contaminant plumes with geoelectrical methods. A case study in urban context. *J Appl Geophys* 75: 738–751.

Venderickx, K., Van Gemert, D., 2000. Geo-electrical survey of masonry for restoration projects. *Internationale Zeitsschrift für Bauinstandsetzen und Baudenkmalpflege* 6 Jahrgang Heft 2 pp 151-172.

Viles, H., Sass, O., Mol, L., 2008. 2D resistivity surveys as a tool for investigating moisture in historic masonry walls. *Proceedings of the International Workshop in Situ Monitoring of Monumental Surfaces* (SMW08) P439-p444 (Tiano, P. & Pardini, C., edts).

Wu, Y., Nakagawa, S., Kneafsey, T. J., Dafflon, B., Hubbard, S., 2017. Electrical and seismic response of saline permafrost soil during freeze - Thaw transition. *J Appl Geophys* 146: 16–26.

Yeh, H. F., Lin, H. I. Wu, C. S., Hsu, K. C., Lee, J. W., Lee, C. H., 2015. Electrical resistivity tomography applied to groundwater aquifer at downstream of Chih-Ben Creek basin, Taiwan. *Environ Earth Sci* 73: 4681-4687.

Yousefi, M. R., Hafizi, M. K., Salari, B., 2017. Gold Exploration Using Induced Polarization and Resistivity Methods in Meyduk-Latala area. *79th EAGE Conference & Exhibition 2017* – Student Programme, Paris, France, 12 – 15 June 2017. DOI: 10.3997/2214-4609.20170 1546.

INDEX